GW01374285

*Nature as Designer: A Botanical Art Study*

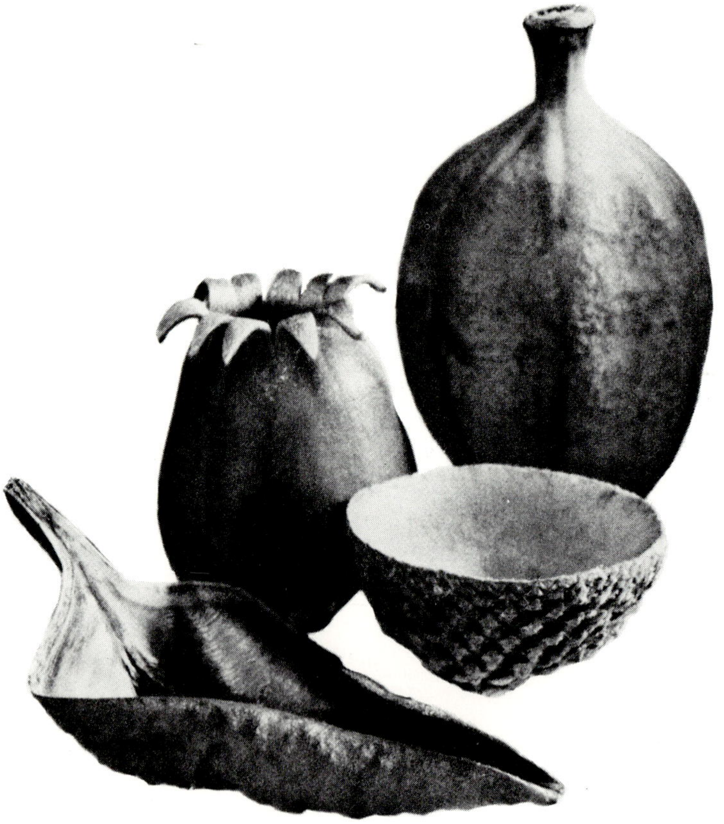

BERTEL BAGER

# *Nature as Designer*

A Botanical Art Study

**VNR** VAN NOSTRAND REINHOLD COMPANY
NEW YORK   CINCINNATI   TORONTO   LONDON   MELBOURNE

Van Nostrand Reinhold Company Regional Offices:
New York, Cincinnati, Chicago, Millbrae, Dallas.

Van Nostrand Reinhold Company International Offices:
London, Toronto, Melbourne.

Library of Congress Catalog Card Number: 66-13731
ISBN 0-442-22017-0

All rights reserved. No part of this work may be reproduced or used in any form or by any means—graphic, electronic, or mechanical, including photocopying, recording, taping, or information storage and retrieval systems—without written permission of the publisher.

Printed in the United States of America
Translated by Albert Read
Originally published in Sweden under the title
»Naturen som formgivare»
by Nordisk Rotogravyr/Stockholm/1961

Published in 1976 by Van Nostrand Reinhold Company
A Division of Litton Educational Publishing, Inc.
450 West 33rd Street, New York, N.Y. 10001

16  15  14  13  12  11  10  9  8  7  6  5

# Foreword by Harry Martinson

How nature works with the motion and resistance of the elements is well illustrated on a windowpane, where frost has executed a fantastic landscape, with palm groves, ferns and laciniated mosses, all elegantly designed by the impact of the currents of warm air in a room against the cold surface of the glass. But why have such patterns as these appeared on the glass, so imitative of living nature that, although we see them a thousand times, they always arouse our wonder?

It may be that there are certain fundamental laws regulating design, laws that are generally valid, in living and inanimate nature and that there is a large, yet limited number of possible shapes among which nature can choose, but no more.

There is strong evidence that this is so. If there were unlimited choice in design, it would be impossible to speak of laws of form; we should have anarchy, an infinite, chaotic multitude of whims of nature, differing from one occasion to another.

How can we then formulate a natural law of design? It cannot possibly be concerned with size, only with form, and must be valid for suns as well as for the tiniest atom.

Let us imagine a sculptor's studio. We will imagine that the sculptor's name is Nature, and that he has been asked to calculate how many different figures can be made of the sphere. When he has done this, he will be given a sphere for every figure he is to carve.

Nature, the sculptress, ponders and calculates. She soon comes to the conclusion that the number of spherical figures must be limited. In any case, it is not greater, the figures must be very much alike. And since all the variations, of whatever kind they may be in a geometrical sense, can be recognized as basically a sphere, the whole question of form very definitely has been delineated.

There is a theory that rounded pottery was first made in imitation of the forms of nature. This theory can hardly stand critical scrutiny. To prove this, let us imagine a number of rounded forms, cherries, bowls, soap bubbles. The closer these approach the sphere, the fewer the variations must be. This applies to art as well as nature. Thus all hand-thrown pottery must generally resemble the round figures in nature. This similarity has nothing to do with imitation, but is due to the rotation of the potter's wheel. The potter's vessel and the "eternal shapes" of pottery are the result of a series of movements forcefully organized into form in a given material.

What this form is to be is determined by the controlled rotation of the potter's wheel, and by his manipulating actions.

For a potter to work from nature modeling a bulb, for example, would be an unnecessarily roundabout way to attain pottery forms with the clay which he could arrive at far more easily by simply making use of the two elements, motion and resistance. The difficult art of the potter consists of using these two elements correctly.

But can we not regard the bulb-shape domes, so common

in Russia, as imitations of nature with a bulb as the prototype? It is possible, of course, but not very likely. And it is still less likely that the bluebell provided the pattern for the bell-founder who cast the first bell. In a bell, man has undoubtedly found the most suitable shape for its function. At the same time it was a natural consequence of rotation and the tools and method of work employed by the bell-founder. This suggests application and communication of quite a different kind and of a far greater scope than would be possible if imitating nature were the only lodestar of art. Nature in her design and man in his art meet somewhere beyond the limits of imitation—in something universally common to both, in the geometric laws valid for the whole universe. Deviations from ideal shapes give rise to variations in form in both art and nature.

Dr. Bertel Bager, who has written, collected and assorted the material for this book over a long period of years, is a man who has long been associated with both art and nature. In addition to his position of chief surgeon at one of the largest hospitals in Sweden, he has made the study of design and nature his avocation. He annually enjoys tours of discovery in the Swedish countryside, searching the grasses and thistles, mosses and lichens, usually finding something worth collecting and meditating upon. Not primarily interested in bright colors, he is fascinated by meadows at the time they have been deserted by butterflies, and by mountain forests, the realm of timeless lichens and mosses. Long after the meadow has lost its summer glory, he seeks the unappreciated—the fruits and seedpods—swaying and rustling in the autumn winds. When winter comes, he stands meditating upon the expressive play of shadows of dead plants on the snow.

What Bager is seeking above all, is the beauty of form in nature. Displayed in many cabinets is his treasured collection of specimens amassed over the years, much like a gallery, showing the fascinating forms of simple things in nature and emphasizing the theme, "Nature as Designer." This volume is a guided tour through part of these treasures but is meant to stimulate interest in form and to please the eye, which is the primary purpose of an illustrated book.

# Introduction

At every step when wandering through woods and fields, one is confronted with beautiful and singular things. The joy of recognition and recollection is felt most keenly when rambling through former childhood haunts. My pleasure never ceases in finding twin-flower runners just where they have always grown among the stumps of trees, to see purple loosestrife as brilliant as ever on the slopes of a summer meadow, to smell again the acrid odor near a fox's lair, or hear the well-known sound of a brook, its waters swelled by melting snow, rushing downward, the thousand memories recalled whenever one revisits the countryside one has known since childhood.

There is a rich reward in the joy of discovery experienced when wandering aimlessly over meadows and wooded hills, or along well-worn paths to beds of blue anemones or lilies-of-the-valley. There is always something new to be found, something never seen before, or something seen in a new light. The way is bordered with question marks. It is not enough merely to see, but to understand all the things, great and small, that nature has to offer. Why is a flower shaped and patterned as it is? Is there any system in its color? Is everything purposeful, according to plan? Is it reasonable to think that all the beauty confronting us exists solely for its own sake and to give us pleasure? What is the fundamental reason for their presence? Such questions crowd upon us leaving many to be answered.

Pleasure in new discoveries is often accompanied by amazement at not having noticed them before.

Using the picture illustrations in this book, I intend to describe the fruits of plants, those parts which bear the seed, and the infinite variety of their forms. Fruit, in everyday speech, means the delicacies summer and autumn bring us: apples, pears, plums, grapes, peaches and the like; it is not these alone I speak of, but all kinds of fruit, even the most insignificant—the botanist's fruits, the final objective of all plants, and of all flowers. One of the first fruit I observed with new awareness was that of the bladder campion. Already slightly acquainted with the plant, I knew its flowers were beautiful white stars with a singularly inflated calyx. One autumn, I discovered a ring of brown spines or teeth in the mouth of the calyx. My curiosity was aroused and, very carefully, I removed the withered remains of the calyx and found a shiny, brown cup. In color, shape, and surface it was vividly reminiscent of pottery, a miniature stoneware vase. I stripped one cup after another of its brown mantle and wondered why such masterpieces should be concealed in withered tissue.

I began to pay more attention to the fruits of plants after this discovery and was amazed at this new world before me. If flowers and fruits are compared aesthetically, flowers are usually given the place of honor, but such comparisons are not always justified. A painting is not necessarily more beautiful than an etching or a piece of sculpture, although you may prefer one form of art to another. Flowers and fruits are both beautiful in their own way. In flowers, colors dominate, with an infinite variety of shades, patterns of dots, lines and patches, a delicate and intricate structure, the singularly beautiful shapes of pistils and stamens. But despite all their glory they are still fragile and transient. Fruits are quite different in their firm and compact composition, their usually simple contours and subdued colors. It is only berries which, when ripe, have bright, attractive colors and which birds and other creatures eat, thereby dispersing the seeds far and wide. The more durable dry fruit seedpods have firm walls, and are storage places for the sensitive seeds of the plant which they must protect during the critical period of growth and until the parent plant is destined to part with them. I was astonished at my own blindness, but I soon noticed that few people had eyes fully opened to the beauty of such fruits. The manner in which individuals experience nature differs from one person to another and it is fascinating to study these differences. One is enamored of sea and water, islands and bays and looks with contempt at the world of the landlubber. Another must combine his love of nature with hunting and fishing. Others most enjoy the grandeur of nature, the wide plains stretching away to the horizon, while still others prefer the silence of the forest with soft carpets of moss or they may long for uninhabited towering mountains. Many are captivated by the interplay of autumn colors—the exuberance of reds, yellows and browns and by the patterns on the ground made by the sun filtering through foliage, and the contrast of light and shade when day is breaking or at twilight. Others find joy in details of nature—catching sight of a bird's nest in the densest of shrubs, realizing the presence of a somnolent lizard basking in the sun on a tree stump of his own color. Many people whose eyes I had thought were open to everything have been astonished about the finds I have made. Professional botanists are, to a certain extent, exceptions, for their work, concerned with the observation of detail is described meticulously in theses and periodicals on flowers, seeds, leaves, and roots. For them, the scientifically significant aspect of form is function—the processes of life, pollination, the dispersal of seeds, hereditary traits and the like. So many books have been published emphasizing the beautiful world of flowers and plants, yet fruits are generally neglected or totally ignored.

In the book, *Läran om det sköna* (Theory of Beauty), by L. Dietrichson, a Norwegian art historian, and published in 1870, there is a chapter on beauty in the vegetable kingdom. The author writes of the beautiful flowers, the vitality of seeds, the geometric severity of tropical vegetation, the beautifully significant grouping of deciduous trees, the spruce forest's solemn aura of mourning in contrast to the pine forest's wildness, etc., a multitude of singular observations. The following passages are especially interesting: "In primitive mosses and lichens the design potential is poor as they develop without much variation or singularity." "The fruits, however, so important to the naturalist, are of smaller significance to the aesthetic."

Although Dietrichson's work was published so long ago, I have a feeling that the same unenlightened words might well be written today.

But the beautiful fruits have not been completely ignored in literature. I have found a German book, *Urform der Kunst*, ("Basic Forms of Art"), by Professor Karl Blossfeldt, which was published in 1928 with a fourth edition appearing in 1948. The illustrative photographs of flowers, leaves, stalks and fruits give an excellent impression of the fascinating forms of these plants and are an inspiration to further study. Many more photographs and illustrations of fruits may be found in botanical textbooks. Another German work, published in 1953, *Baumeister Natur* (Nature the Master Builder), by Dr. Horst Janus, has many beautiful illustrations of the vegetable kingdom, including fruits. There is also a series of French books from the Flammarion publishing house called, *Collection le montreur d'images*. The separate titles are: *Découvertes* (Discovered), *La vie cachée de fleurs* (The Hidden Life of Flowers), *De la fleur à la graine* (Seed Flowers), *Les bourgeons s'ouvrent* (The Buds Open), and *Ramures* (Branches).

One reason why we so seldom see garden plants with fruit is that the flowers are usually cut off and the plants are removed when they have withered because they are unsightly and untidy in the garden. Perennial plants, particularly bulbs, weaken if left to ripen, and their flowers the following year would be less beautiful. However, I would advise flower lovers to allow at least some flowers in the garden to keep their fruit, or seedpods, so they may become acquainted with them. Perhaps some of the pictures in this book will open their eyes to the beauty latent in the plants. When the potter, thousands of years ago, began molding his clay into pitchers and bowls, he was making things for everyday use, using, perhaps, a gourd or the curved shell of a coconut as his model. It would be interesting to know how far the creations of nature, particularly fruits, were used as prototypes and sources of inspiration for early potters. There is no doubt that in both decoration and form, purely artistic impulses appeared very early. Two present-day potters, Axel Salto of Denmark, and Toini Muona of Finland, give credit to nature for their inspiration. When one considers what the Lord has done as a potter, it must be admitted that no purer source of inspiration exists. Wherever one turns, one is confronted with His masterpieces, and in museums, fossils all bear witness to a wonderful, now extinct, vegetable world, one that should direct the mind to the study of art as do fragments of ceramics unearthed in ancient Greece, China and elsewhere. The discovery of various new fruits aroused my interest in other details of plant structure. Early buds, for example, in their simple construction and precise form, are very reminiscent of fruits. Their vaulted envelopes protect the rudimentary leaves or flowers, just as the walls of a pod protect the seeds. The horse-chestnut and rhubarb buds provide good examples of the enormous vital force in plants and it is delightful to follow their development under the warm spring sun. Many buds droop at first, and only gradually turn up to face the sun when the flower is ready to open its petals to their full beauty. Generally speaking, spring is the prime time for buds, but it is interesting to examine many autumn buds, the catkins on hazel shrubs, the resinous shoots of pine, the green and black tips of the ash branch. Linnaeus, in his account of a journey in Scania, complained that botanists paid too little attention to buds. "Botanical literature," he wrote, "lacks a full description of the buds of trees, and how finely they were made by the Creator to protect the tiny plant with its leaves and blossoms which lie safely in the bud until the following year. Had botanists studied these buds, they probably would have found the answers to many obscure problems."

Leaves are responsible for the green of summer landscapes, and there is abundant green there—green forests, many shades of green in fields of fresh new vegetables, green marshes and shore grasses, thickets, and velvet lawns. Chlorophyll molecular variations create the varying tones of green. Varying greens suggest the thought of the single individual leaf and the beauty of its detail. In botany, the shapes of leaves are described exactly using adjectives necessary for classification—oval, spatulate, orbicular, sagittate, lobed, pinnatifid, etc.

But behind these coldly descriptive expressions are realities with most beautiful and interesting variations. The dandelion leaf with its sharp-toothed edges, the elegant spear-points of black bindweed and willow leaves (sagittate), the green shields of the nasturtium and the marsh pennywort, the rolled edges of wild rosemary and Crystal tea leaves, different in color and structure on the upper and lower surface, the filigree-like leaves of horsetails, the widely different leaves of conifers and sedums, the decorative trailers of Virginia creeper, ivy and the morning glory —to mention only a few of the multitude of beautiful leaves nature can display. We also have the different patterns on the leaf surfaces, the lines or network of veins and nerves, the fine shades of clover leaves, the blotches on the leaves of the spotted orchid, the sporangia in regular rows on the backs of fern fronds, the stinging hairs of nettles and the smooth surface of the leaves of aquatic plants.

Nor should old leaves be neglected. Many age and wither in beauty, beside the spectacular autumn show of color change with which we are so familiar. Some thistles twist their withering leaves into elegant whorls. The great lobed leaves of the autumn anemone shrivel in a curious way, with different shades of color on either side. It is fascinating to follow the process of decomposition even further, and see how a leaf dries to a fine gray network. The same process can be studied in many fruits, in those of the henbane, various poppies and Scottish bluebells. Generally speaking, it is safe to say that the beauty of withered plants is too often overlooked. The skeletons of trees and plants can be singularly beautiful. Among herbaceous plants, those belonging to the family Compositae deserve special attention. Both cornflowers and red knapweeds become everlasting flowers when the florets have withered and dropped away and the seeds have ripened and left, dispersed by the winds. The empty baskets are off-white and shine like silver, and may well be used for decoration in winter when flowers are scarce. The white stars remaining when the seeds of groundsel have sailed away on their parasols are beautiful against a dark background. The enormous seedheads of the giant hogweed keep their beauty indoors almost indefinitely, as do branches of the greater water-plantain, growing in loose pyramidal panicles. Many plants, when they wither, become tough and hardy, and stand out as decorative silhouettes against the snow. The composition becomes even more perfect when their contours are drawn in shadows by the sun. Consider grass stalks with their empty panicles, tussocks of sedge in bogs, wild chervil with its airy umbrels, the coarse-growing docks, almost ghost-like in the twilight, and the tiny round fruits of bedstraws on stalks as fine as hair. This is the time of the year I call the etcher's time. Even the branches of trees and shrubs appear as if etched against bright snow and sky.

There is a miniature vegetable kingdom, the mosses and lichens, where species and families are not so well known as the more common flowering plants. I feel sure that most people know nothing more about the soft carpet they walk on in the woods than that it is moss, and they know still less about lichens. There are not many authentically native names for these plants, most of them have simply been assigned professionally by botanists. But a knowledge of names, even scientific ones, does not seem essential to me in this connection. It is quite impossible for a layman to remember the many families and species of this world in miniature. But that need not interfere with his enjoyment of the beautiful and singular shapes to be seen. Someone once called lichens Flora's stepchildren, for the glorious colors she has lavished on other plants are denied to lichens. I do not agree with this for although it is true that their colors are seldom gaudy, their delicate gray, brown, and green shades are most charmingly combined, and one never tires of their forms. Some pictures of these cryptogams will be found in this book.

Another group of cryptogams, many with familiar old native names, are attractive to crowds of people out-of-doors in autumn; they are the fungi. For many people, the brightly colored varieties are a delight to see, but it is highly probable that the thought of subsequent culinary enjoyment is uppermost in their minds to judge from the ruthless way they will destroy unknown species, regardless of how beautiful they may be. This is both thoughtless and deplorable. Let the unknown fungi stand in peace, study them carefully and simply enjoy the beauty of another of Nature's works of art. Many of the hymenomycetes (toadstools) resemble stoneware bowls, although the bowl is upside-down at the foot of the fungus. Consider the ruffled agaric, for example. When it first appears, the crown looks like a simple bucket, but within a few days Nature improves the design until the final goal is reached—a large open bowl with scale-like decoration in brown shades. Other species stand on the ground like bowls or funnels; morels and the bowl fungi proper, are some of the most beautiful with caps of pale, yellowish-brown on the surface and a bright red underneath.

When botany is one's hobby, and the scientific names of plants are met regularly, an explanation of such names is often desired. Many of them are derived from ancient Latin, Greek and Arabic, and the old names of the same or similar plants. Linnaeus was a genius in the art of naming plants, and he seems to have been a master of Latin also. During his long journeys, he took note of native names. Like the scientific names, some of these are of ancient date, so old that their origin is lost in the dawn of civilization. All our trees have such names, short and pithy—oak, pine, beech, spruce, maple, birch, aspen, ash, rowan, etc., provincial variations are almost non-existent. It is otherwise interesting to see how varied the names of plants may be in different sections of a country, how one characteristic of the plant determines the name in one area and another feature is chosen elsewhere. Frequently the use made of some part of it gives the plant its name. The more one tries to analyze these names, the more interesting this branch of botany becomes.

In closing this introduction, I must stress that this book is a picture book. The illustrations should, as far as possible, speak for themselves. Let me quote Cajsa Warg, who, in the foreword of her famous eighteenth century cook book wrote of her style in a manner which expresses my own feelings: "As far as the style is concerned, I hope it is intelligible though simple, without the least embellishment of a borrowed quill, for, as I have always found a dish unsavoury in which spices and condiments have been used to excess, so should I also deem the flowers of elocution spread over my descriptions, the real value of which is found only in their clarity and truth."

My hesitation in publishing the results of my hobby was overcome, perhaps, by an egoistic wish to experience the joys of sharing my hobby with others. I mentioned the joy of recognition and discovery one experiences in rambles through field and forest, but it is also pleasant to show later, what one has found, particularly since I believe that my discoveries are new to many people. I recall a story told by Anna Maria Roos in 1908, in an article entitled "The Child as Artist." She told the story to show how far the inability to use one's eyes may go. "A hunting party sat down to breakfast in a glade, and an old gamekeeper, who had accompanied hunting parties for decades, bent down and picked a twig of flowering wood anemone. "What beautiful balsam," he said. The hunters laughed and corrected him. He had never noticed the flower growing in the woods. In vain they tried to help him recollect that he must have seen such flowers everywhere in his sixty years of life in the woods as a gamekeeper. "This may seem incredible," says Miss Roos, "but it is literally true, that even a long life in woods and fields cannot teach a person to use his eyes, and it is highly probable that everyone must be trained to see. It is, in reality, a disadvantage for people to live without the habit of looking and the ability to notice. Many moments of delight, and the possibilities of enjoying beauty are missed if they have not trained their ability to see." I entirely agree with Anna Maria Roos' conclusions on her incredible story, and if this book can help to open people's eyes to all that is beautiful in nature, I will be gratified and happy.

1 · MONTAGE

# Orientation

Every year I go through my school herbarium in order not to forget the names I once learned. It is always annoying to recognize a plant and yet be unable to recollect its name. When looking through my collection, I find that the plants are, generally speaking, well pressed and neatly arranged, but still a herbarium seldom provides any great degree of aesthetic pleasure, except perhaps, by way of the associations it may have with the living specimen in its natural environment. One category is particularly neglected in herbaria, and that is fruit. This is probably because fruit cannot be pressed satisfactorily. When my interest in fruits was first aroused, the problem of how to store them arose. It was necessary to arrange a "fructarium" which would always be available for study. At first I kept my finds in boxes divided into compartments, but soon realized that they could not be viewed to the best advantage and the boxes were difficult to handle. The next step was to mount the single fruit or seedcase, effectively. As a base I used sheets of cardboard of different colors, cut into convenient shapes and sizes, and sometimes thin pieces of wood. The objects were attached with sealing-wax or glue (page 14). Later I used modeling rubber which can be rolled into a suitable thickness or molded into pedestals for different kinds of fruit. This material provides good support for the objects, and is especially suitable when they are to be photographed, for it holds them firmly in any position. The fruit should be kept in cupboards, preferably with glass doors to keep out the dust. They may also be stored in glass showcases exactly as ceramics are exhibited. The photograph on the back of the book jacket shows a collection arranged in this manner.

Only dry fruits, naturally, can be collected and kept in cupboards or showcases. Soft, fleshy fruits and berries, and unripe, not yet ripened seedpods, must be kept in hardening baths of alcohol or formalin, but they will not retain their beauty as dry fruits do; their colors often fade or disappear completely and their surfaces become macerated. But some, rose-hips and water-lily fruits, for example, may remain beautiful for a long while.

The collection of material on which my book is based was not planned in advance. Every time I go out walking I find something worth taking home to study. It may have been my interest in pottery that first drew my attention to dry fruit and seedpods, which are similar in many ways to the creations of artists working in clay and with the potter's wheel. Gradually, my enthusiasm was aroused for all kinds of remarkable details in the vegetable kingdom, and when pictures had to be selected the arrangement became more systematic. I found it desirable to include buds and leaves and needles, to give examples of interesting methods for the dispersal of seeds and fruits, to exhibit decorative seedcases, to show the beautiful details of mosses and lichens. My greatest difficulty lay in selecting specimens from among all the wonderful things nature so lavishly has to offer. It would be simple to fill several books with pictures. More than 50 of the photographs are of wild plants I have collected in my walks through northern Södermanland and in Roslagen in Sweden, about 25 are from gardens, and about as many are from botanical gardens; about ten are from museum collections, and a few were collected on a trip through Portugal.

Now that my book is to be published, I have become increasingly aware of the many people to whom I owe a debt of gratitude. First I must thank Mr. Harry Martinson who encouraged me in my work and who energetically insisted upon my presenting these pictures to the public. I must also thank him for his great kindness in writing the foreword to my book. Professor R. Florin allowed me to freely use the specimens of the Bergianska Botanical Gardens in Stockholm, and together with Mr. E. Söderberg, helped me to identify several plants. Professor E. Hultén, chief of the Botanical section of the Museum of Natural History, also offered me a wide choice of specimens of the museum's extensive collection. He and his assistants, Mr. S. Ahlner, Mr. T. Hasselrot, and Dr. H. Persson of the Paleobotanic department authenticated a number of the plants included in the book. Professor Emeritus C. Skottsberg has given me a beautiful exotic plant, as well as good advice; his interest in my work was a great source of encouragement. I extend my sincere thanks to all these scientists. I also owe a debt of gratitude to Dr. Anna Nordenskjold, who helped me with the literature study. My wife and family have followed my work with stimulating interest in spite of my having grossly neglected them during certain critical phases of my work. At the same time I must pay tribute to their patience when I found it necessary to litter the house with all the "rubbish" I brought home from my field trips.

The following people were of great assistance in the photography of specimens: Miss Elsie Karlsson, assistant photographer at the X-ray department of Mörby Hospital, Mrs. Märta Claréus of the Nordic Museum, and for a brief period, Mr. N. L. Sääf, of Esselte Foto. Most of the photographs were taken by Mr. Erik Lundqvist of Nordisk Rotogravyr. I wish to thank all of them for the great interest they took in this effort, and for the understanding they demonstrated in my intentions.

# Contents

The first number in each line is that of the picture in the book.
The numbers before the + or − symbols indicate the degree of
magnification or reduction.

The major part of the photographs were made by Erik Lundqvist,
the others by Märta Claréus (numbers 4, 7, 31, 66, 71, 72, 94, 99, 100, 134, 173,
184, 185, 186, 189, 191),
Elsie Karlsson (85),
Lennart Sääf (26, 48, 70, 102, 111, 124, 125, 141, 154, 163, 170, 172),
and Bertel Bager (87).
The color photographs on the jacket are by Erik Lundqvist.

| | | | |
|---|---|---|---|
| 1 | Montage . . . . . . . . . . . . . . . . 14 | 25 | Datura stramonium, Thorn apple, 3+ . . . . . 39 |
| 2 | Quercus robur, English oak, 3—4+ . . . . . 20 | 26 | Monotropa hypopitys, Yellow bird's nest . . . . 40 |
| 3 | Ditto, 3—4+ . . . . . . . . . . . . . 21 | 27 | Ditto, 10+ . . . . . . . . . . . . . 41 |
| 4 | Ditto, 9—10+ . . . . . . . . . . . . 21 | 28 | Argemone grandiflora, Prickly poppy, 3+ . . . 42 |
| 5 | Papaver, Poppy, various species, 2—3+ . . . 23 | 29 | Lecythis pisonis, actual size . . . . . . . . . 43 |
| 6 | Ditto, 3—4+ . . . . . . . . . . . . . 22 | 30 | Lagenaria vulgaris, Gourd, 1/2— . . . . . . . 44 |
| 7 | Ditto, 2—3+ . . . . . . . . . . . . . 24 | 31 | Cucurbita, a species of gourd, somewhat reduced 45 |
| 8 | Ditto, 2—3+ . . . . . . . . . . . . . 24 | 32 | Pithecoctenium echinatum, actual size . . . . . 46 |
| 9 | Ditto, 2—3+ . . . . . . . . . . . . . 25 | 33 | Banksia grandiflora, somewhat enlarged . . . . 47 |
| 10 | Meconopsis cambrica, Welsh poppy, 5+ . . . . 26 | 34 | Casuarina equisetifolia, Beefwood, 6+ . . . . . 48 |
| 11 | Hyoscyamus niger, Henbane, 1/2+ . . . . . 27 | 35 | Zeyhera montana, actual size . . . . . . . . 49 |
| 12 | Ditto, 7—8+ . . . . . . . . . . . . . 26 | 36 | Couratari pulchra, actual size . . . . . . . . 50 |
| 13 | Lilium Martagon, Turk's-cap lily, 2+ . . . . . 28 | 37 | Couratari macrosperma, somewhat enlarged . . . 51 |
| 14 | Ditto, actual size . . . . . . . . . . . . 28 | 38 | Cistus ladaniferus, Gum cistus, 6+ . . . . . . 52 |
| 15 | Aquilegia vulgaris, Columbine, 3—4+ . . . . 29 | 39 | Paris quadrifolia, Herb Paris, 4+ . . . . . . 53 |
| 16 | Primula veris, Cowslip, 4+ . . . . . . . . 30 | 40 | Species of rose, 2—3+ . . . . . . . . . . 55 |
| 17 | Melandrium album, White campion, 5+ . . . 31 | 41 | Ditto, 10+ . . . . . . . . . . . . . 54 |
| 18 | Caryophyllaceae, Clove, slightly enlarged . . . 32 | 42 | Ditto, 4+ . . . . . . . . . . . . . 56 |
| 19 | Saxifraga granulata, Meadow saxifrage, 5+ . . . 33 | 43 | Ditto, 3+ . . . . . . . . . . . . . 57 |
| 20 | Aristolochia labiosa, Dutchman's pipe, 4+ . . . 34 | 44 | Nigella damascena, Love-in-a-mist, 4+ . . . . 58 |
| 21 | Eucalyptus globulus, Blue gum tree, 3—4+ . . . 35 | 45 | Ditto, 4+ . . . . . . . . . . . . . 59 |
| 22 | Eccremocarpus scaber, Glory flower, 3—4+ . . 37 | 46 | Nigella hispanica, 5+ . . . . . . . . . . 60 |
| 23 | Ditto, 4—5+ . . . . . . . . . . . . . 36 | 47 | Nigella sativa, 3+ . . . . . . . . . . . 61 |
| 24 | Ledum palustre, Crystal tea, 3—4+ . . . . . 38 | 48 | Nymphaea candida, White water-lily, 3—4+ . . . 62 |

17

| | | |
|---|---|---|
| 49 | Nuphar luteum, Yellow water-lily, 3–4+ | 63 |
| 50 | Ditto, 4+ | 64 |
| 51 | Blumenbachia spiralis, 3–4+ | 65 |
| 52 | Iris pseudacorus, Yellow iris, 3–4+ | 67 |
| 53 | Ditto, 3–4+ | 66 |
| 54 | Eschscholtzia californica, California poppy, somewhat enlarged | 68 |
| 55 | Parnassia palustris, Grass of Parnassus, 4+ | 69 |
| 56 | Ditto, 4+ | 69 |
| 57 | Acer pseudoplatanus, Sycamore maple, 3–4+ | 70 |
| 58 | Acer platanoides, Norway maple, 6–7+ | 71 |
| 59 | Lonicera symphoricarpus, Honeysuckle, 4–5+ | 72 |
| 60 | Ditto, greatly enlarged | 73 |
| 61 | Anthriscus silvestris, Wild chervil, 3+ | 74 |
| 62 | Ditto, 6+ | 74 |
| 63 | Filipendula ulmaria, Queen-of-the-meadow, 5+ | 75 |
| 64 | Sparganium simplex, Unbranched bur-reed, 5+ | 76 |
| 65 | Nicandra physaloides, Apple-of-Peru, 4+ | 77 |
| 66 | Ditto, 2+ | 77 |
| 67 | Hibiscus esculentus, 2+ | 78 |
| 68 | Hibiscus trionum, 3+ | 79 |
| 69 | Lonicera caprifolium, Honeysuckle, 4+ | 80 |
| 70 | Ditto, 4+ | 81 |
| 71 | Molucella laevis, 7+ | 83 |
| 72 | Ditto, 5+ | 82 |
| 73 | Malope trifida, 4+ | 84 |
| 74 | Seed vessels, 4+ | 85 |
| 75 | Calla palustris, Water-arum, 8+ | 86 |
| 76 | Caiophora hibiscifolia, 3+ | 87 |
| 77 | Caiophora cernua, 3+ | 87 |
| 78 | Campanula persicifolia, Large bellflower, 3–4+ | 88 |
| 79 | Campanula carpathica, Tussock bellflower, 5+ | 89 |
| 80 | Salix caprea, Goat willow, 4+ | 90 |
| 81 | Ditto, 6–7+ | 90 |
| 82 | Ditto, 1/2+ | 91 |
| 83 | Ditto, 4+ | 91 |
| 84 | Taraxacum vulgare, Dandelion, 4+ | 92 |
| 85 | Ditto, somewhat enlarged | 93 |
| 86 | Ditto, 2+ | 93 |
| 87 | Ditto, 1/2– | 93 |
| 88 | Asclepias syriaca, Common milkweed, actual size | 94 |
| 89 | Ditto, 2+ | 95 |
| 90 | Cynanchum vincetoxicum, Black swallow-wort, actual size | 94 |
| 91 | Tragopogon pratensis, Goat's-beard, 4–5+ | 96 |
| 92 | Scabiosa stellata, 3+ | 97 |
| 93 | Anemone tomentosa, 3+ | 98 |
| 94 | Ditto, actual size | 98 |
| 95 | Epilobium angustifolium, Great willow-herb, 2+ | 99 |
| 96 | Thlaspi arvense, Penny cress, 2+ | 100 |
| 97 | Ditto, 6+ | 100 |
| 98 | Viola tricolor, Wild pansy, 4+ | 101 |
| 99 | Geranium silvaticum, Crane's-bill, 4+ | 102 |
| 100 | Ditto, 5+ | 103 |
| 101 | Erodium gruinalis, Stork's-bill, 2+ | 104 |
| 102 | Ditto, 5+ | 105 |
| 103 | Erodium cicutarium, Hemlock stork's-bill, 7+ | 105 |
| 104 | Pinus silvestris, Pine, 2+ | 107 |
| 105 | Ditto, 4–5+ | 106 |
| 106 | Ditto, 2+ | 107 |
| 107 | Ditto, 2+ | 108 |
| 108 | Ditto, 2+ | 108 |
| 109 | Ditto, 3+ | 109 |
| 110 | Picea abies, Norway spruce, actual size | 111 |
| 111 | Ditto, 4+ | 110 |
| 112 | Ditto, actual size | 111 |
| 113 | Araucaria columnaris, New Caledonian pine, 2+ | 112 |
| 114 | Juniperus communis, Juniper, 7–8+ | 113 |
| 115 | Carlina acanthifolia, 3–4+ | 114 |
| 116 | Cirsium lanceolatum, Spear thistle, 3–4+ | 115 |
| 117 | Helianthus annuus, Sunflower, 2+ | 116 |
| 118 | Doronicum caucasicum, Leopard's-bane, 4+ | 117 |
| 119 | Atractylis cancellata, 2+ | 118 |
| 120 | Ditto, 5+ | 119 |
| 121 | Rudbeckia purpurea, Daisy, 3+ | 120 |
| 122 | Matricaria inodora, Scentless mayweed, 3+ | 120 |
| 123 | Rudbeckia lanceolata, 5+ | 121 |
| 124 | Arctostaphylos uva-ursi, Bearberry, 7+ | 122 |
| 125 | Ditto, 5+ | 123 |
| 126 | Pyrola secunda, English wintergreen, 2 1/2+ | 124 |
| 127 | Prunella vulgaris, Self-heal, 6+ | 125 |
| 128 | Salvia viridis, Sage, 2–3+ | 126 |
| 129 | Oenothera graveolens, Evening-primrose, 2+ | 127 |
| 130 | Trematolobelia macrostachys, 4+ | 128 |
| 131 | Lobelia gloria-montis, 3+ | 129 |
| 132 | Astragalus falcatus, actual size | 130 |
| 133 | Ditto, actual size | 130 |
| 134 | Lunaria annua, Honesty, 1 1/2+ | 131 |
| 135 | Solanum dulcamara, European bittersweet | 132 |
| 136 | Isatis tinctoria, Dyers woad, 4+ | 133 |
| 137 | Corylus avellana, Filbert, 4+ | 135 |
| 138 | Ditto, 6+ | 134 |
| 139 | Ditto, 4+ | 134 |
| 140 | Platanus acerifolia, Plane tree, 4–5+ | 136 |
| 141 | Centaurea cyanus, Cornflower, 5–6+ | 137 |
| 142 | Aesculus hippocastanum, Horse-chestnut | 139 |

| | | |
|---|---|---|
| 143 | Ditto, 3+ . . . . . . . . . . . . | 138 |
| 144 | Tulipa, Tulip, 2+ . . . . . . . . | 140 |
| 145 | Ditto, 2—3+ . . . . . . . . . . | 140 |
| 146 | Bidens tripartita, Three-cleft bur-marigold, 6+ | 141 |
| 147 | Geum rivale, Water avens, 4+ . . . . . . . | 142 |
| 148 | Geum urbanum, Herb bennet, 4+ . . . . . | 142 |
| 149 | Arctium (Lappa) tomentosa, Burdock, 2—3+ . . | 143 |
| 150 | Ditto, 2—3+ . . . . . . . . . | 143 |
| 151 | Medicago obscura, 5—6+ . . . . . . . . | 144 |
| 152 | Medicago sativa, Alfalfa, 2—3+ . . . . . | 145 |
| 153 | Medicago disciformis, 5—6+ . . . . . . | 145 |
| 154 | Populus tremula, Poplar, 4—5+ . . . . . | 146 |
| 155 | Alchemilla vulgaris, Lady's mantle, 3+ . . . | 147 |
| 156 | Iris germanica, actual size . . . . . . . | 149 |
| 157 | Ditto, 3+ . . . . . . . . . . . . | 148 |
| 158 | Drosera rotundifolia, Round-leaved sundew, 5+ | 150 |
| 159 | Ditto, 4+ . . . . . . . . . . . . | 151 |
| 160 | Ditto, 4+ . . . . . . . . . . . . | 151 |
| 161 | Aristolochia durior, Dutchman's pipe, 2+ . . . | 153 |
| 162 | Ditto, actual size . . . . . . . . | 152 |
| 163 | Saponaria officinalis, Bouncing Bet, 6+ . . . | 154 |
| 164 | Anemone silvestris, Snowdrop anemone, somewhat enlarged . . . . . . . . . . | 155 |
| 165 | Allium schoenoprasum, Chives, 3+ . . . . | 156 |
| 166 | Pithecolobium dulce, somewhat enlarged . . . | 157 |
| 167 | Cytisus laburnum, Laburnum, 3—4+ . . . . | 158 |
| 168 | Lycopodium clavatum, Common clubmoss, 3+ . | 159 |
| 169 | Equisetum arvense, Common horsetail, 2 1/2+ . | 160 |
| 170 | Equisetum limosum, Water horsetail, 1/2+ . . | 161 |
| 171 | Equisetum arvense . . . . . . . . . | 161 |
| 172 | Equisetum limosum . . . . . . . . . | 162 |
| 173 | Equisetum arvense . . . . . . . . . | 163 |
| 174 | Phleum pratense and Cynosurus cristatus, Timothy, 4+ . . . . . . . . . . . . | 165 |
| 175 | Cynosurus cristatus, Crested dog's tail, 4+ . . | 164 |
| 176 | Carex vesicaria, Bladder-sedge, actual size . . . | 166 |
| 177 | Ditto, 3—4+ . . . . . . . . . . | 166 |
| 178 | Cladonia gracilis var. chordalis, 2—3+ . . . . | 167 |
| 179 | Cladonia gracilis, 2+ . . . . . . . . | 168 |
| 180 | Cladonia pyxidata, 2+ . . . . . . . . | 168 |
| 181 | Hylocomium splendens, 2—3+ . . . . . . | 169 |
| 182 | Ptilium crista-castrensis, 2—3+ . . . . . | 169 |
| 183 | Polytrichum formosum, 5+ . . . . . . . | 170 |
| 184 | Solanum tuberosum, Potato, 3+ . . . . . | 171 |
| 185 | Paeonia officinalis, Peony, 3+ . . . . . | 172 |
| 186 | Ditto, 5+ . . . . . . . . . . . | 173 |
| 187 | Paeonia suffruticosa, 3+ . . . . . . . | 173 |
| 188 | Antirrhinum majus, Snapdragon, 2+ . . . . | 174 |
| 189 | Ditto, 4+ . . . . . . . . . . . | 174 |
| 190 | Scorpiurus sulcatus, 3—4+ . . . . . . | 175 |
| 191 | Citrus aurantium, Orange, 2+ . . . . . . | 176 |

2 · QUERCUS ROBUR

A comparison of the solid acorn falling heavily from the oak, and the light, winged seed of the pine, carried for miles by the wind, shows clearly how differently nature has treated the problem of seeds. Both the oak and the pine are large, strong trees, yet an acorn is a thousand times as heavy as the pine seed.

Man has long been interested in acorns, and this interest has been manifested in many ways. Swine grunt happily as they grub for acorns in oak woods, gamekeepers gather acorns for pheasants to eat in severe winters, in lean times acorns are roasted to satisfy a craving for coffee, and old-time apothecaries made decoctions of them. Artists and designers have often used acorns as models, and we find them in ornamental reliefs, or stylized in etchings, on old pewter lids or in the work of silversmiths. The acorn is also an unusually beautiful and pleasant fruit, both to look at and to touch. Close scrutiny will reveal that they differ greatly in shape and size. There are slender, pointed acorns, and large, almost barrel-shaped ones, some with a hint of a waist. Every oak seems to have its own type. The acorn caps are just as interesting. They are perfect, small living bowls, reminiscent of beautifully finished stoneware and are also varied in design. Some are low and flat, others enclose almost half of the acorn, while still others are slightly oblique. Finally, it is worth the trouble to follow the development of the acorn from its earliest beginning, when the tip—the remains of the pistil—begins to emerge from the fruit cup, until the shiny brown acorn, often striped with dark lines, is fully grown.

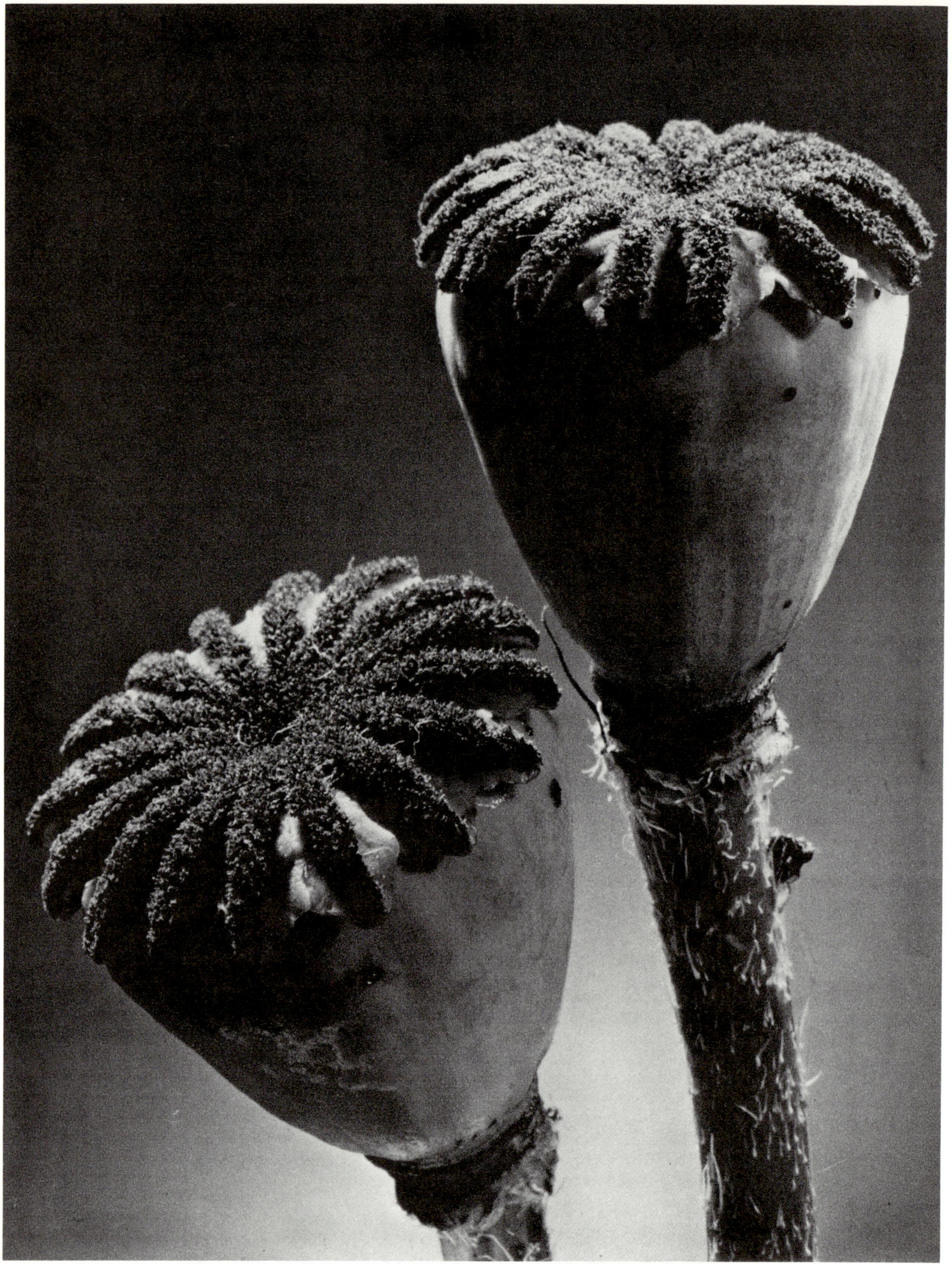

5 · PAPAVER

The gigantic flowers of the red poppy, cultivated in our gardens, and the wild poppy growing by the thousand in cornfields, have the same incomparably glorious color. Seed catalogs enthusiastically list the many varieties. But the singular and beautiful, often splendid, seedcases are not usually mentioned. The strong pistil alone, is a notable sight, with its stigma spreading over the broad ovary like a violet-black starfish. Poppy buds hang like heavy drops, but the stem straightens up when the flower begins to open, and becomes stiff, almost wooden.

The seedcase opens in a curious way. A German author, Bartsch, noticed a strange analogy as he wrote of the poppy in a book called, *Botany for Ladies*, published in 1810, "the fruits open at the sides like the embrasures of a man-of-war." The seeds are dispersed by the wind, which rocks the

23

stiff stems to and fro, scattering the seed in all directions through the openings. If one of these seedcases is turned upside-down, hundreds of seeds run out, like sugar from a sieve.

Anyone who has enjoyed seeing a glassblower at work will get the impression that the poppy fruit on its stem is like a glass bowl on the end of a blow-pipe, perfectly formed. The seedpods are delicately colored, usually pale yellowish-brown, or terracotta, and not infrequently with shades of violet. The poppy has a fruit one never tires of looking at, and I advise those who grow poppies in their gardens to allow some seedpods to ripen by not cutting off all the withered flowers. He may then cut them to fill vases when flowers are scarce. The glorious flowers certainly brighten our gardens for a few summer days, but the seedpods are everlasting.

The illustrations are of cultivated species of poppy. In late autumn, when wind and rain have almost completed their work of disintegration, airy skeletons of seedcases, like those on the opposite page, can be found.

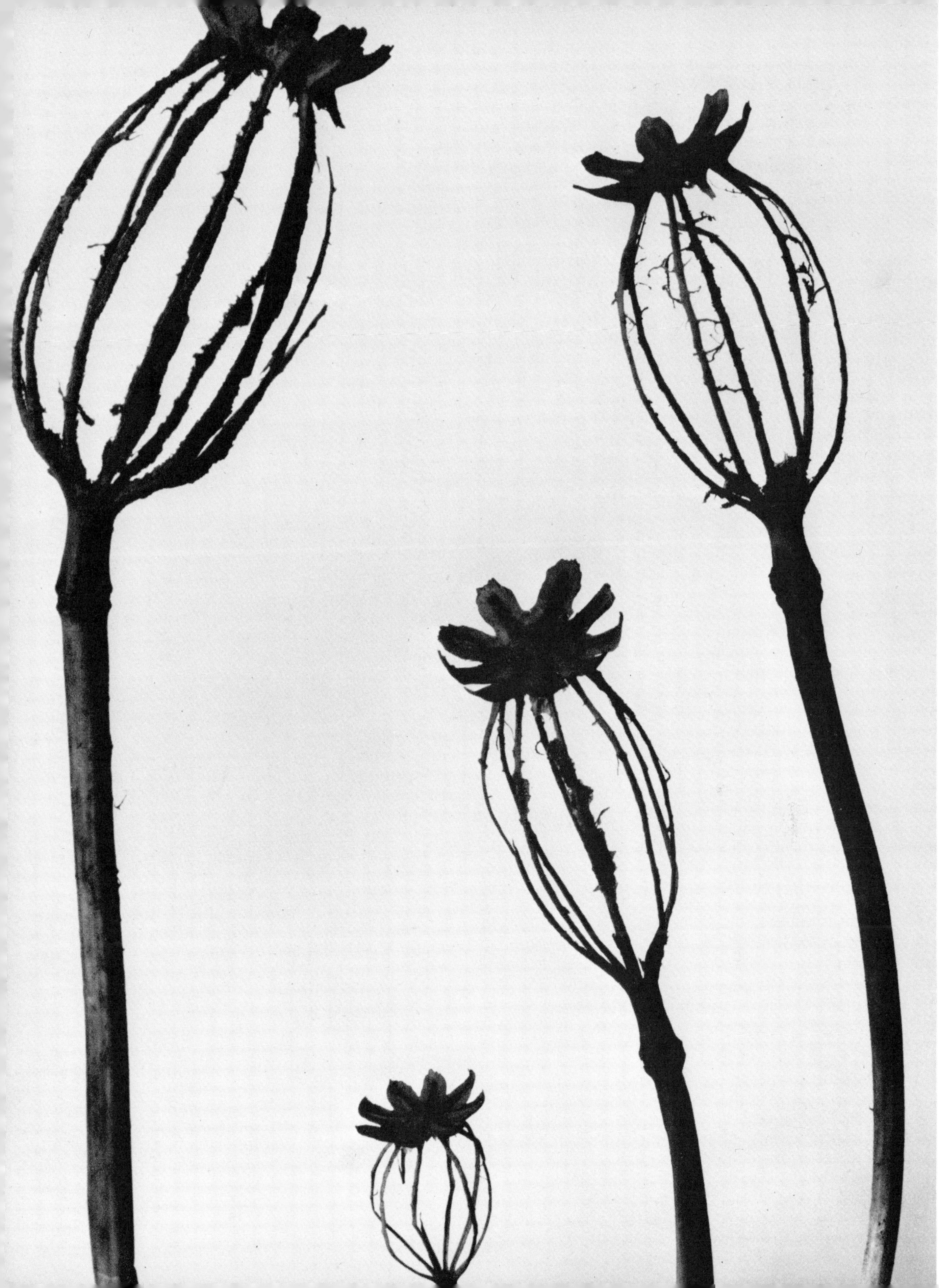

It is easy to see from the seedcases that *Meconopsis* belongs to the poppy family. Several species grow in the Himalayas and in China, but only one in Europe. I found the species illustrated here, *Meconopsis cambrica* (the Welsh poppy), in the beautiful old botanical garden at Leyden, growing in a border near the empty pedestal of a bust of Linnaeus, destroyed during the war and lying in pieces under a table in the garden cottage. The elongated seed capsule terminates in a short pillar or knob, supported by five arches. Longitudinal ridges decorate the surface of the capsule, and between them is a fine network of nerves. The Latin name of the species means Moonface, which, when one considers the round yellow flower, is a very apt description.

10 · MECONOPSIS CAMBRICA

12

11 · HYOSCYAMUS NIGER

Those who examine seedcases and fruits with the eyes of a potter must feel happy when they see the beautiful cups of the henbane, standing in a long row along one side of the strong stem. The fruits are built like vases with a distended base, and a slender waist from which the upper half widens into five pointed lobes. What one sees is the calyx, pale yellowish-brown in color, with a network of nerves. The leaves and flowers wither early in autumn, but the stems and seedcases remain for a long time, frequently throughout the winter. When spring comes, the tissue of the mantle, after long exposure to the weather, gradually falls away, leaving a fine filigree cup. This cup is, however, only a protective covering for an inner seedcase fashioned like a small casket on a wide base, with a curved lid. This casket is so like a type of low, bronze, Chinese bowl known as Yu, of the Shang dynasty, that it is difficult to dismiss the idea that the first artist who made such a vessel must have been inspired by the henbane's seed capsule. When autumn storms tear at the stiff stems, the lid loosens and falls off, and the rough seeds, which look like wild strawberries in the photograph, follow. The capsule is a veritable cornucopia, for each plant may produce up to 8000 seeds. Henbane is poisonous, and men have long made use of the poison. It is very likely that the hallucinations which the Oracle at Delphi needed to perform her duties as a soothsayer were produced by poison from the henbane. Murderers and primitive medicine-men made decoctions from the plant. It is still in the Pharmacopoeia as the source of hyoscine, a sedative. The beautiful fruits symbolize a medicine glass as well as the poison cup.

13 · LILIUM MARTAGON

The Turk's cap lily *(Lilium martagon)* is a common garden flower, often found running wild in old gardens. It is grown for the sake of its blossoms, but when they have withered, the plant is forgotten until the following year, when it again opens its beautiful corollas. But it should not be so quickly forgotten, for one can enjoy it after the flowers have faded. The fruit is very decorative, and it is interesting to follow their development and see how they rise from the pendant position of the flowers until they stand stiffly upright. If a seed capsule is cut off at the main stem, and placed with its stem upward, it resembles one of the elegant flasks seen in laboratories. The fruit of the Turk's cap lily is very suitable for decorative use in vases, but they should be cut before they ripen, otherwise the capsules will soon open, and they are much more beautiful when closed. Such "cut fruit" will last almost indefinitely.

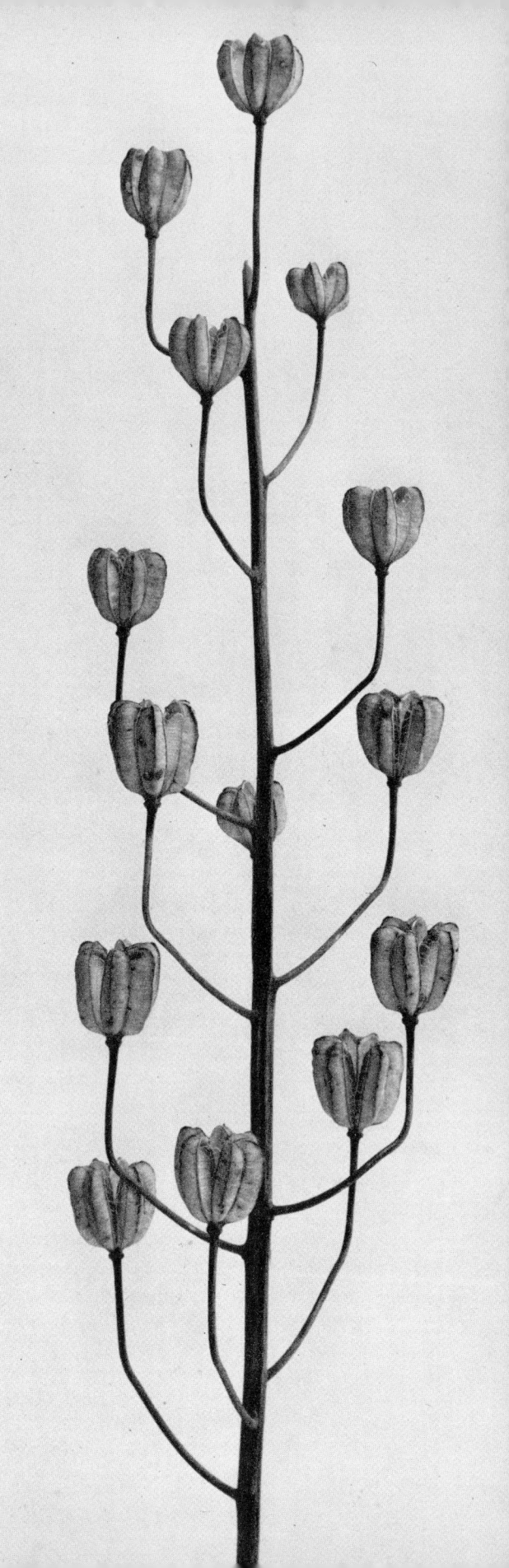

15 · AQUILEGIA VULGARIS

The columbine is a very popular garden flower often advertised in seed catalogs, and continually appearing in new varieties—blue, violet, white, yellow, red, pink, and variegated; with single and double flowers, with long or short spurs. But I have not seen the fruit mentioned except in books on botany, and then only in scientific terms: "fruit, five follicles, spurred." And yet they are attractive from a purely aesthetic aspect. The illustration is of the fruit of the common columbine, the violet columbine with spurs like the cornucopia; it is seldom found wild but is very common in gardens where it will grow profusely. It probably originated in a monastery garden. This specimen was picked in spring. The long, narrow follicles are empty; the tiny black seeds, which filled them in autumn, have been dispersed by the wind. The veins of the follicles form a fine network, sometimes with transverse lines, like the underside of a snake. These beautiful markings can be seen best in spring after the plant has been exposed to wind, rain, and frost through the autumn and winter.

16 · PRIMULA VERIS

The separate flowers of the cowslip do not all come out at the same time. The first to flower have already withered and stand upright, while others are still hanging in bud or flower. Out in the fields the withered plants are not noticed, for new flowers attract the eye, and conceal the dead plants. But in late summer, those who look will find the withered cowslips brown and rather ugly. It is worth while, however, to look a little closer. Just after flowering, when the petals have fallen, the calyx dominates. The five leaves are joined together and form a hairy pitcher, terminating in pointed lobes. It is green and decorative, with faintly marked ridges but it withers gradually and becomes a wrinkled envelope for the fruit, which can be seen from above. Down, more or less deep in the calyx, is a dark surface face surrounded by a ring of brown teeth. This is the opening of the seedcase, and with the help of a magnifying glass the seeds can be seen inside it. If the ugly calyx is removed carefully, the naked seedpod, a beautiful little pale-brown pitcher, can be seen. The illustration shows four fruits, two with and two without the calyx.

17 · MELANDRIUM ALBUM

The *Melandrium album*, a beautiful flower, has other more attractive names, such as white candle pink or field lantern. But I find the fruit even more attractive. It was one of the first to open my eyes to the beautiful world of fruits. Actually, the plant is a rather untidy one at fruit-setting. The calyx, which is inflated like a balloon, with distinct vertical lines, turns brownish-black and threadbare, and completely envelops the seedcase; it is only at the tip that the mouth of the seedcase with its toothed edge, can be seen. If the withered tissues are removed carefully with tweezers, the naked fruit can be seen in all its glory. It is a small, perfect ceramic-like bowl, nut-brown in color, and tapering upward from the bulbous base to the top, which opens in a ten-pointed star. If the seedcases are turned downward, numerous seeds run out; if there is a magnifying glass handy, examine the seeds carefully, they have a curiously marked surface, with tiny grains in curved rows.

18 · SPECIES OF CARYOPHYLLACEAE

Species of caryophyllaceae make a display resembling a ceramic collection. The lower row of seedcases shows how the shapes vary, but are still, in principle, very similar to their related species. They all belong to the Caryophyllaceae, the Pink family, and include *Viscaria, Dianthus, Silene* and *Lychnis*. Most of these can usually be found in botanical gardens. I cannot remember the names of all species, and perhaps it does not matter very much, for botanical classification is not important here. The series is a variation on a theme.

In the upper row are specimens of *Melandrium, Agrostemma* and some seedcases belonging to other families, a *Gentiana* and a *Nemesia*, for example.

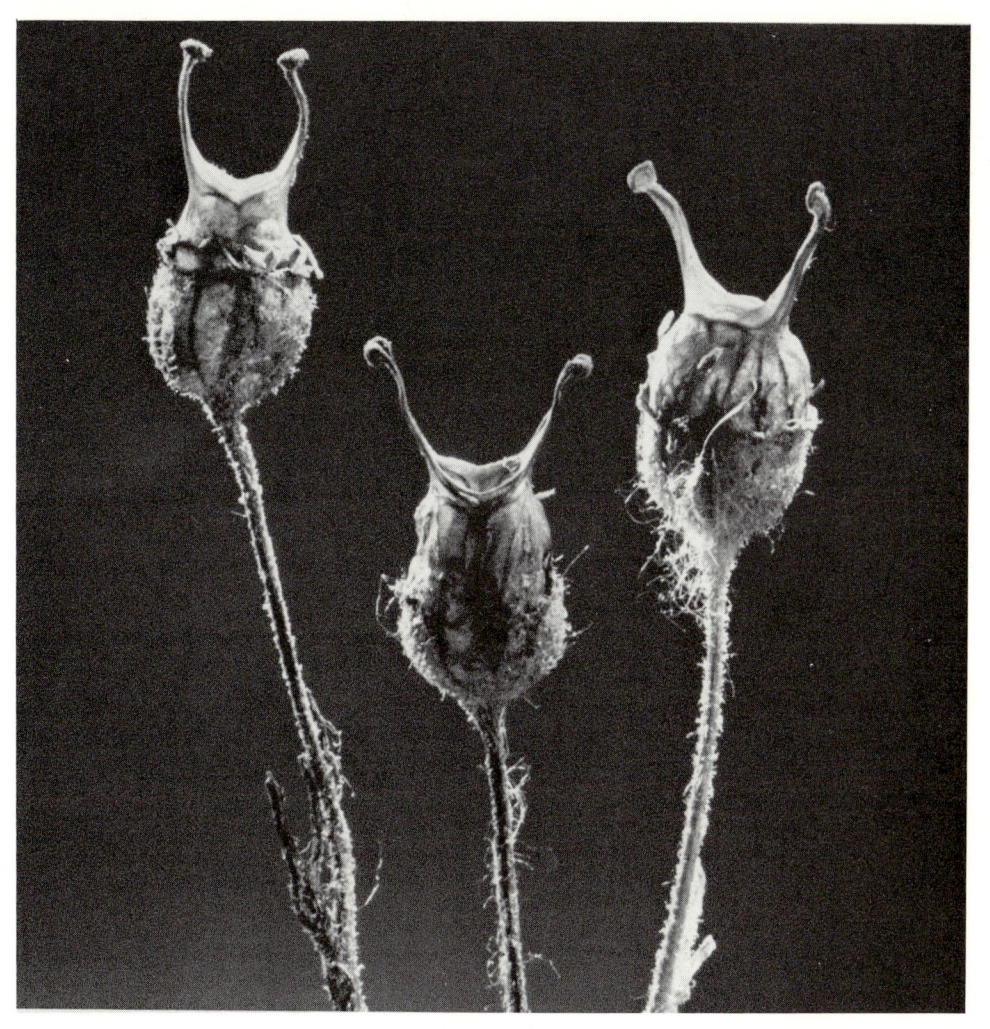

19 · SAXIFRAGA GRANULATA

The meadow saxifrage is a charming flower. It appears faithfully and punctually, year after year, with a wealth of blossoms. They often grow in dense clusters and seen against the light, when the stems are half grown, there is a bewitching lustre over the entire plant, due to the dense glandular hairs. The rosettes of leaves survive the winter on rocky slopes, and spring to life as soon as the snow has melted. The fruit is a globular capsule with two long appendages, which are the hardened remains of the pistil. Between them is an opening, through which the seeds are shaken by the wind. The seedcase looks like an old drinking-cup of a type used in the fifteenth and sixteenth centuries, a carved, wooden ale pitcher, with two tall handles.

20 · ARISTOLOCHIA LABIOSA

The Dutchman's pipe, whose liana-like stem is shown elsewhere in this book, belongs to a family with many tropical species. The fruit illustrated above was given to me by a friend, who picked it in an old park at Hammamet, south of Tunis. It belongs to the species *Aristolochia labiosa*. The capsule is divided into five follicles, separated from each other. The finely toothed edges look as if they have been kept together with zipper fasteners. The whole capsule looks like an elegant lantern, swaying in the wind hanging by six fine threads, one for each follicle. A decorative little knob puts the finishing touch at the bottom. It is somewhat reminiscent of the fruit of Crystal tea.

34

21 · EUCALYPTUS GLOBULUS

I found the seed capsules illustrated here in the Douro Valley in Portugal, under some stately trees with narrow, hanging leaves, and scaling bark which gives the trunk a mottled appearance. It was a group of gum trees, *Eucalyptus globulus*, native to Australia, but cultivated now in many parts of America and the Mediterranean region, chiefly because of its ability to dry up the soil discouraging mosquitoes from living in the area, and also for its durable timber. When the flowers open, the calyx is thrown off as a small cap. The fruits are strongly built, as hard as wood and with a rough surface. They are a warm brown, except for the plane surface of the cone, which is bluish-gray. The scent of eucalyptus oil can be noticed when two fruits are rubbed together. There are many different species of *Eucalyptus*, and I have seen some in the beautiful old botanical garden in the university city of Coimbra in Portugal. The fruit of one species differed from those illustrated here; they were smaller and rounded, like half an acorn. It must be a wonderful sight to see the birds around flowering gum trees. I have read that in Australia, the blossoms are pollinated by honey-loving parrots, in Chile by hummingbirds, and in South Africa by honey-birds.

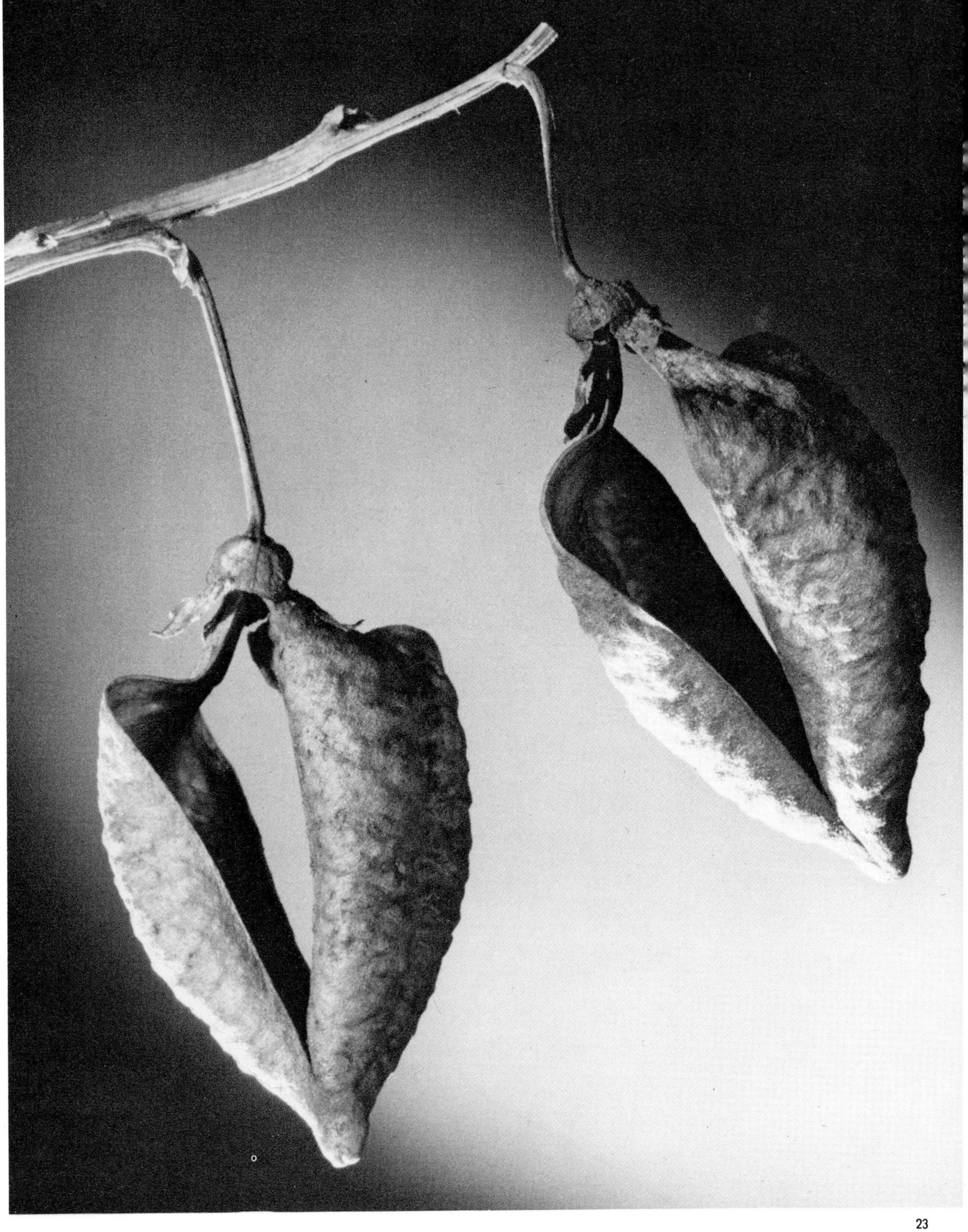

23

22 · ECCREMOCARPUS SCABER

*Eccremocarpus scaber* is a climbing plant, which grows wild in Chile. It has glossy, orange-red flowers, hanging in graceful racemes, and curious fruit. When closed, the fruit is a brown capsule with a bulging base which tapers into a conical tip. When it is ripe it splits into two halves, but not entirely, for the two halves remain joined together at the base and tip. The seedcase then looks something like two hands joined in prayer. The small winged seeds, miniature elm seeds, are ejected through the apertures. It is easy to separate the two halves and we see that each half looks like a tiny earthenware dish; the smooth interior, the brown color, and rough exterior strengthen this impression.

24 · LEDUM PALUSTRE

The dense stands of Crystal tea *(Ledum palustre)* on marshy ground in the woods always remind me of virgin country. I can recognize the acrid, but not unpleasant smell at a great distance. Our forefathers used the tender shoots in place of hops for brewing ale, and I suspect that our beer could hardly stand comparison with the rich mead, spiced with Crystal tea, which they drank. The rust-brown underside of the leaves, and the young shoots of the same color, contrast singularly with the otherwise glossy, dark green leaves. In early summer, the white flowers shine in the dark woods, then follows an intermediate period with unripe fruit, and finally, the bushy panicle with brown, hanging seedpods. Some of the pods have been removed in the picture, so that the others might appear more clearly. They split open from the base into five parts, and expel their many, fusiform seeds. The pistil projects like a bird's bill and, combined with the fruit, gives the impression of a hovering hummingbird.

25 · DATURA STRAMONIUM

The thorn apple *(Datura stramonium)* contains a poisonous substance, which has long been used in the treatment of asthma. Even if it were not poisonous, it would hardly tempt our vegetarian animals, for the leaves are prickly, and the fruit, too, is armed with needle-sharp spines. The wonderful bowls which Chinese potters fashioned are as much a delight to the hand as to the eye, but the splendid thorn apples cannot be handled at all. Yet they have inspired potters. I am reminded of the Danish artist, Salto, who in a book he wrote, illustrated a piece of work he named the demoniacal vase or the Devil's own vase which must have been modeled on the thorn apple.

26 · MONOTROPA HYPOPITYS

The yellow bird's nest *(Monotropa hypopitys)* is a mysterious parasitic plant. The pale yellow color, due to an absence of chlorophyll, arouses our curiosity; the contrast with the green darkness of the woods is striking. A close scrutiny of the plant reveals much of interest. The term *hypopitys* indicates that it grows under pine trees, but it is just as much at home under spruce, and even in deciduous woods. The leaves of the stem are reduced to small scales. The drooping inflorescence rises from its soft, mossy bed, and becomes erect after fertilization. The young stems, which are somewhat similar in color and consistency to asparagus—indeed, the German name of the plant is *Fichtespargel*—are transformed during the summer into hard brown sticks. Flowering specimens are frequently seen growing with old, withered colonies. The flowers form a dense raceme toward the top of the stalk, each with a short pedicel, which continues to grow after the flowers have withered. The seedcases retain the pale yellow color of the flowers during the summer, but become darker in autumn. The pistils, with their flattened stigmas, project above the fruits, giving them the appearance of narrow-necked vases, or old-fashioned decanters. Finally, the pistils fall off, and the seedcases open longitudinally.

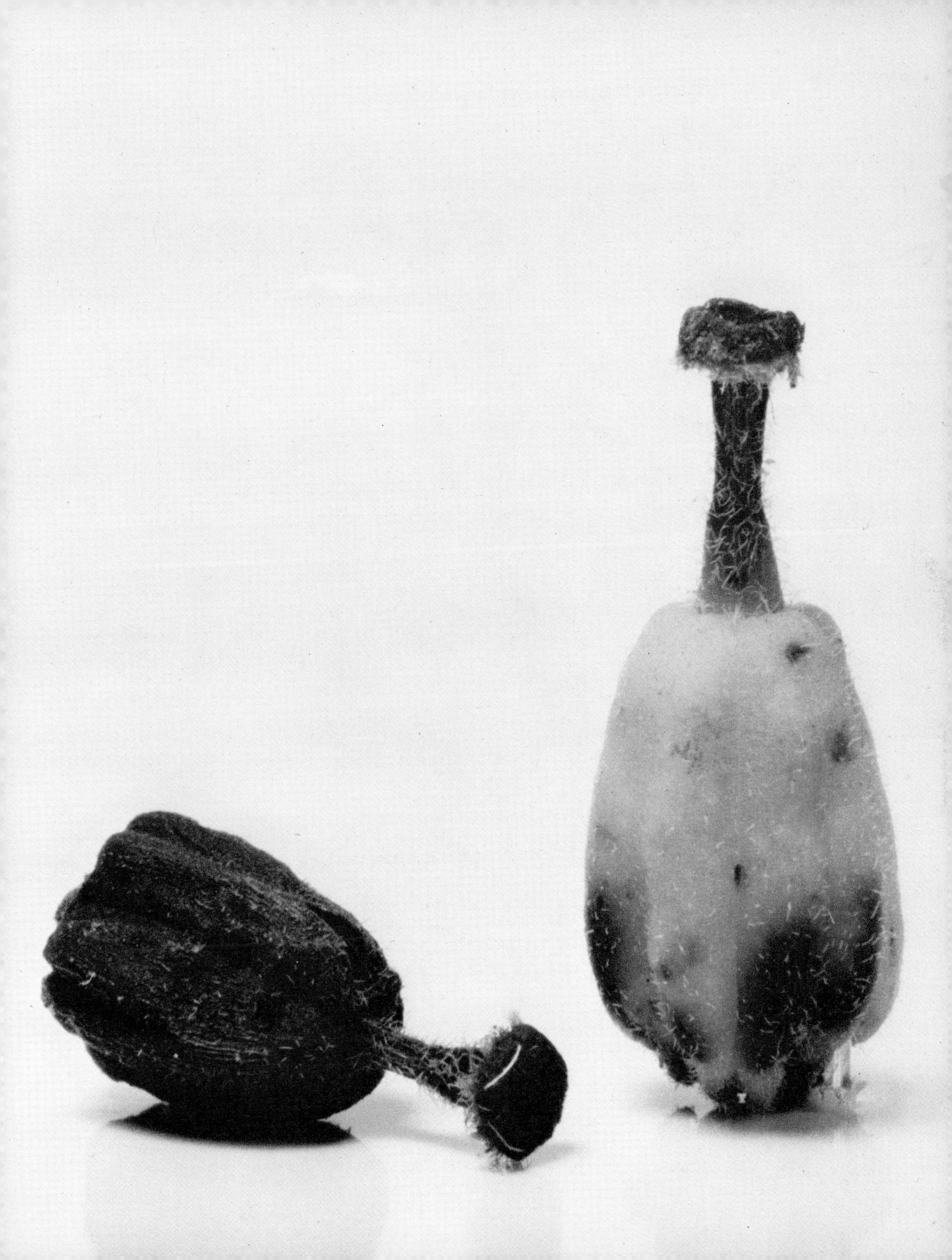

28 · ARGEMONE GRANDIFLORA

*Argemone grandiflora* is closely related to the poppy although this is difficult to believe if the seedcases shown above are compared with those shown on page 24, with their rounded contours and smooth surface. *Argemone* species are sometimes called "prickly poppy," or are known by a more unpleasant name, "the Devil's fig." And no one will envy him this diet. In contrast, nature here has built a vault of five graceful arches, as perfect as the Gothic architecture in our cathedrals.

In the tropical rain forests of South America, grows a family of plants of several species, known as *Lecythis*. Most of the species are tall trees with unusually shaped flowers—the stamens are congregated in an elongated bowl extending from one side like an umbrella over the actual flower—and striking fruit. The species illustrated here, *Lecythis pisonis*, is from the botanical garden in Rio de Janeiro, and the fruit is in the Museum of Natural History in Stockholm. It is, as a Latin description says, "Magnitudine capitis infantis," as big as a baby's head. It is astonishingly similar to an earthenware pitcher; the walls are thick, almost as hard as stoneware, and the lid is conical, with decorative depressions. Another member of the family is well-known for its delicious fruit which we call the Brazil nut *(Bertholletia excelsa)*.

29 · LECYTHIS PISONIS

30 · LAGENARIA VULGARIS

Gourds are mentioned in two places in the Bible. When Jonah sat in the burning sun of the city of Nineveh, the Lord prepared a gourd to create shade over Jonah's head, and Elijah, in Gilgal, when there was famine in the land, sent sons of the prophets to gather wild gourds, which they shred into a pot. These gourds seem to have been poisonous, for Elijah, by adding meal, removed all harm from the pot. Scholars cannot agree on the identity of these plants. There seems to be no doubt that the term "gourd" is derived from the Latin *cucurbita.* The illustration of the gourds shown here is of museum specimens labeled *Lagenaria vulgaris.* This refers only to the two outer specimens for the beautiful fruit in the center has no label, but is most likely closely related to the other two. The right-hand specimen is the natural shape of the gourd, the one on the left has been given its shape artificially by tying it during growth. The gourd is an annual climbing plant from the tropical regions of the Old World, where it was often grown for its fruit which was used as bottles, bowls for pipes, rattles, etc. The fruit is really a gigantic berry with a hard shell. The fruit illustrated here is reduced to half the actual size. The similarity of the left-hand fruit to a modern ceramic vase is striking, and it has understandably been used as a prototype by potters. The picture opposite is of a gourd grown in many varieties.

31 · CUCURBITA

32 · PITHECOCTENIUM ECHINATUM

In the tropical forests of Brazil is a liana, whose scientific name is *Pithecoctenium echinatum*. The first word means monkey-comb, and the second indicates the spinosity. The plant has hanging fruit, shown here slightly reduced, with a dense covering of small sharp spines. The fruit opens from the bottom to release the large winged seeds, which when ripe, are loose within the fruit and layered like the leaves of a book, to slip out easily. The two halves of the fruit form boat-like bowls, pale-brown and glossy inside.

There is a family of plants called *Banksia*, which grows in arid parts of Australia. It includes both trees and shrubs with leathery, evergreen leaves. The fruit is very unusual. That shown here *(B. grandiflora)*, from the botanical garden in Sydney, is slightly reduced. The seed capsules project and open like mussels to release the seeds.

33 · BANKSIA GRANDIFLORA

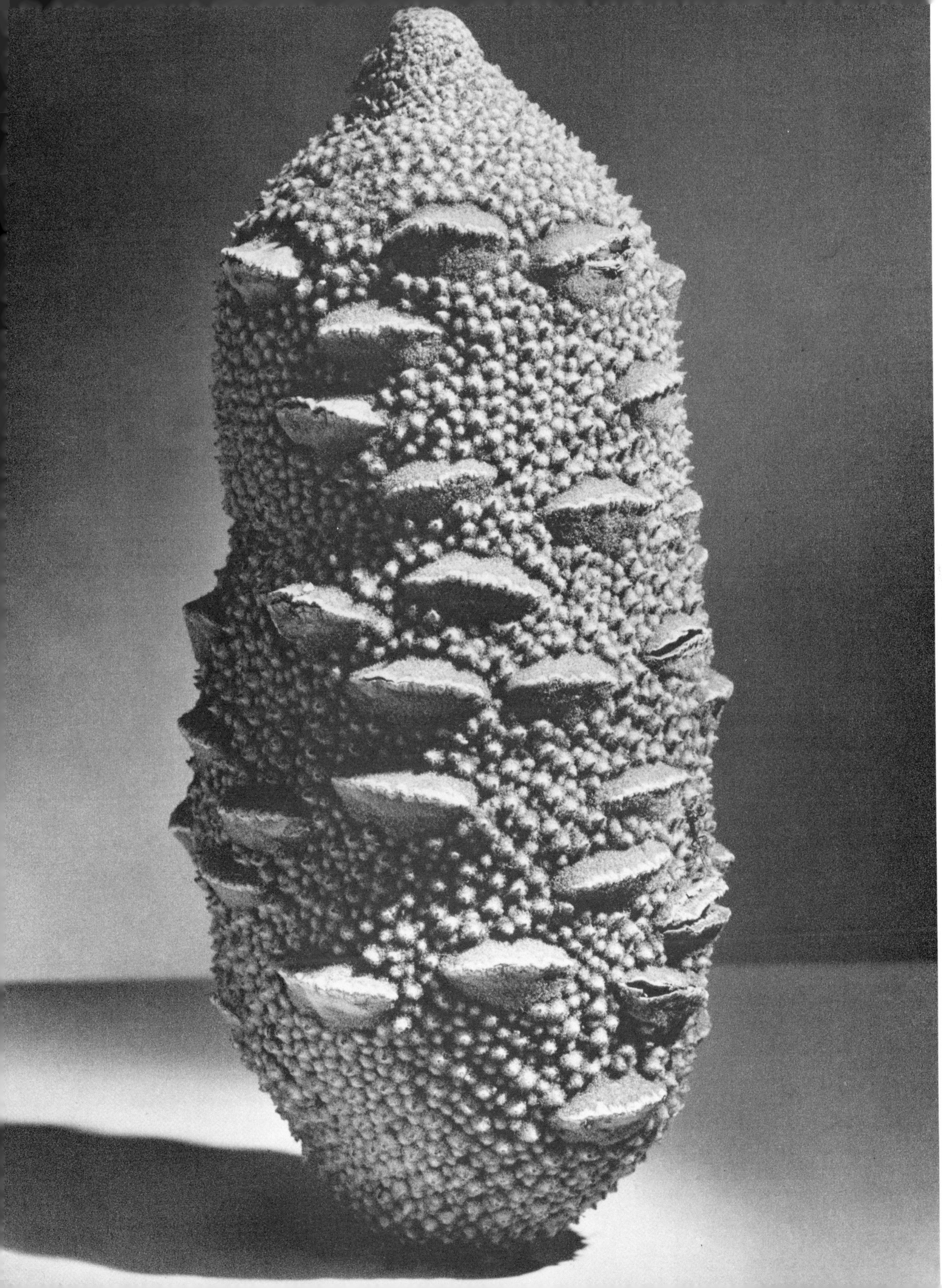

34 · CASUARINA EQUISETIFOLIA

The singular fruit in this photograph was picked many years ago from a tree with the Latin name, *Casuarina equisetifolia*. This, and several trees closely related to it, frequently form large woods in Australia or on the neighboring islands, and are similar in some ways to conifers and horsetails, hence the species name, *equisetifolia*. They all have very hard timber, called beef-wood in Australia, from which the aborigines are said to have made their war clubs. The appearance of the fruit suggests that there might be some symbolism in their choice of wood. With their short stems they are just like clubs, and they are hard and woody, like pine cones. The winged seeds fall from the hollows arranged in long rows on the fruit.

35 · ZEYHERA MONTANA

*Zeyhera montana* is a shrub or small tree of Brazil, growing six feet or more in height. A beautiful inflorescence shoots up from the top, and the five-lobed leaves are green above but brownish-gray underneath. The fruit is shown here actual size. The surface is very rough. Brazilians, who speak Portuguese, call the fruit Bolsa de Paster, which means shepherd's satchel or pouch.

49

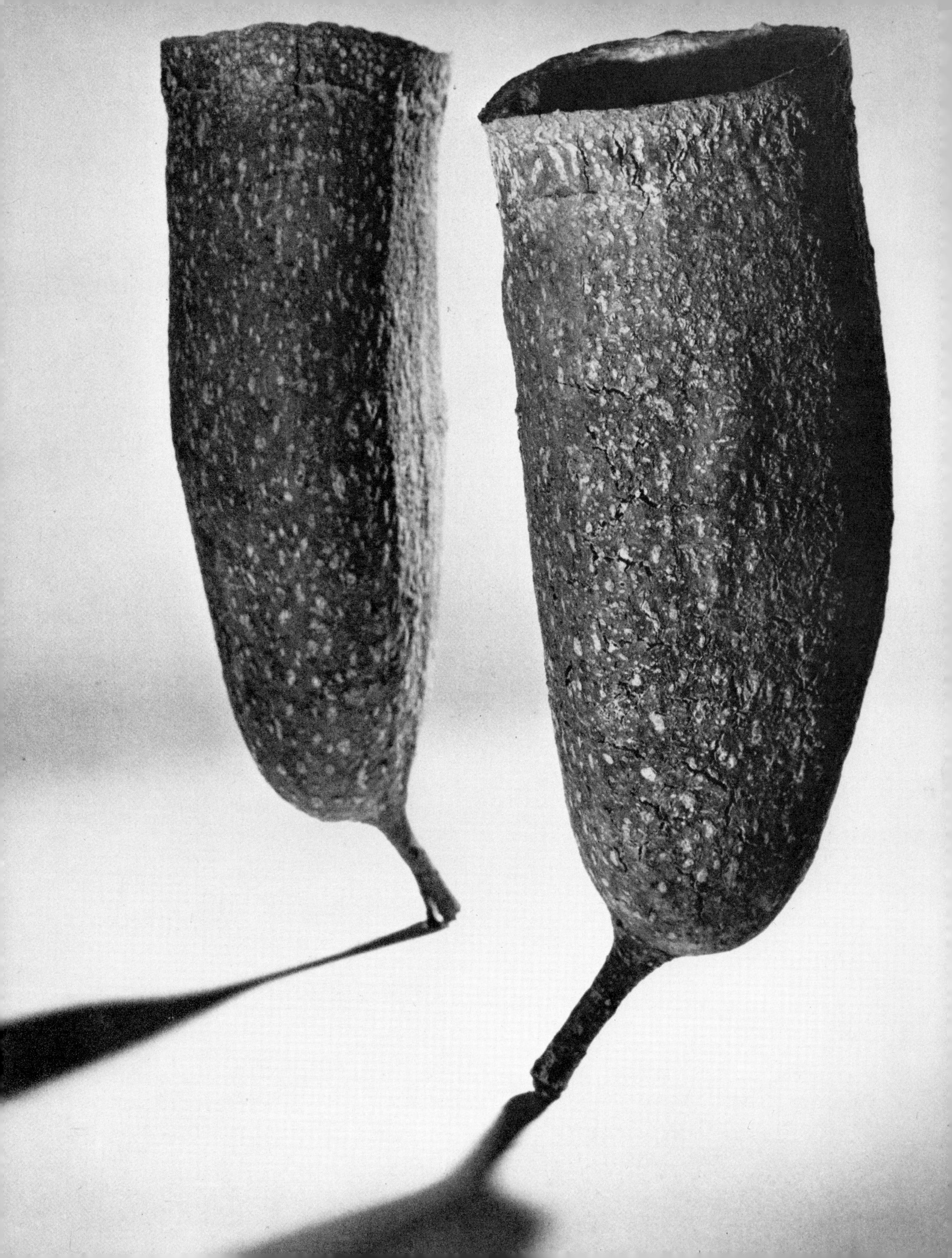

37 · COURATARI MACROSPERMA

These specimens are of the beautiful fruits of two species of *Couratari* from Brazil. The smaller species, the fruit of which is shown opposite, is called *Couratari pulchra*. I do not know whether *pulchra*, which means beautiful, refers to the fruit or the flowers, which, according to descriptions in floras are beautiful shades of yellowish-brown and wine-red. I assume that the adjective refers to the flower, for fruits are usually neglected aesthetically, but *pulchra* would not be out of place as a description of this fruit. The other species is called *macrosperma*, which indicates that the seeds are large. The greater part of the fruit consists of a wide wing-edge. The seeds of both species are as hard as wood, goblet-shape and very much like pottery. Both have lids with knobs in the middle, but in this specimen they are unfortunately lost. The single fruit is shown one quarter actual size, the others with a rather eccentric foot, are slightly magnified. The trees are straight and tall, attaining a height of 100 to 130 feet.

36 · COURATARI PULCHRA

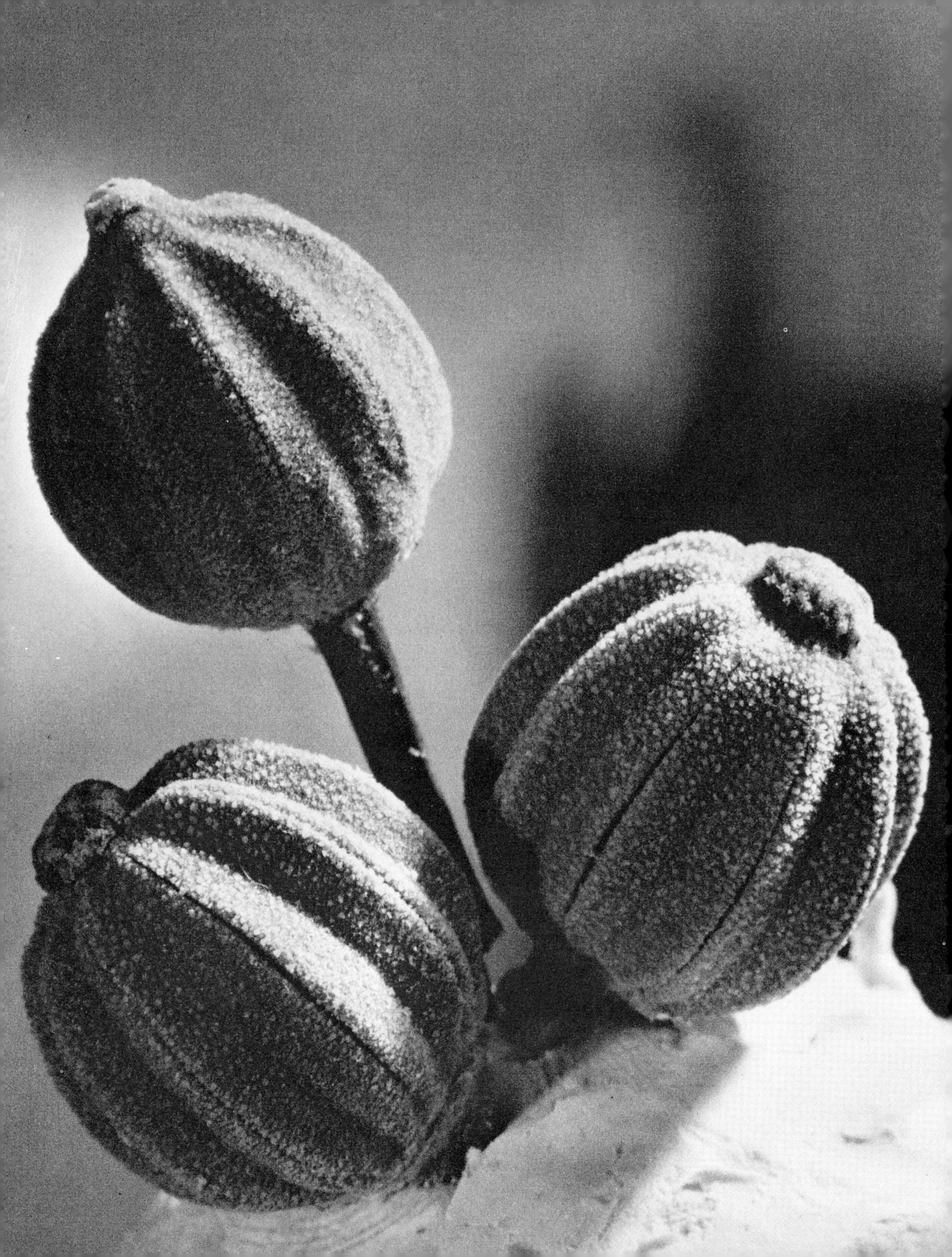

39 · PARIS QUADRIFOLIA

On his way north from the south coast of Portugal, the traveller passes a high plateau before reaching the fertile plains. On both sides of the highway, which winds over the plateau, is a dense growth of shrubs, about five feet high. I recognized a few of the *genista* species with yellow flowers, and then I found a shrub with a strong scent of balsam, and fruit some of which is seen on the opposite page. They are like melons, but no larger than small cherries. The shrub was *Cistus ladaniferus*. A fluid secreted by the fine glandular hairs gives the scent of balsam. This fluid solidifies in the rosin called ladanum, used widely in pharmacy. The hard, regular fruit is divided into sections, each containing numerous small seeds enclosed in a curious latticework.

The species name of herb Paris *(Paris quadrifolia)* refers to the four leaves of the plant, which grow in a ring at the bottom of the flower stem. The origin of the family name, *Paris*, is unknown. Perhaps the similarity of the beautiful berry to an apple made the botanist who named the plant think of the apple which Paris gave to the most beautiful of the three goddesses, Hera, Pallas Athene and Aphrodite. The entire plant can defend its right to a place among beautiful flowers. It is delicately built, with a tall, thin stem; the leaves are beautifully veined, the flower, a curious shape with green, pointed sepals, long, needle-like stamens with yellowish-white anthers in the middle and a spherical, brownish-red pistil. The ripe, blue-back berry rests on a violet star, not unlike a starfish with the rays drooping over the withered sepals.

38 · CISTUS LADANIFERUS

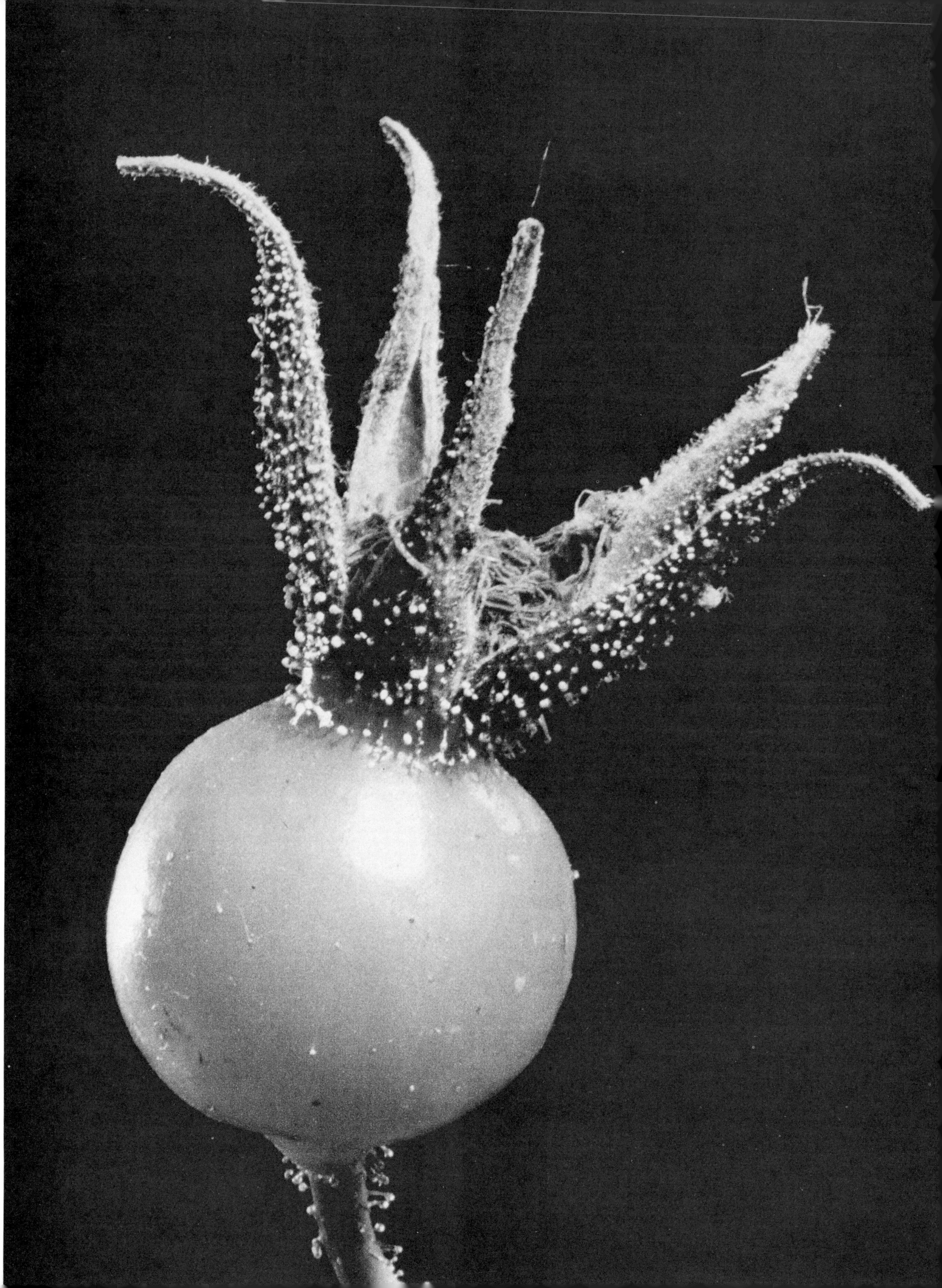

40 · SPECIES OF ROSE

In the introduction, I made some observations about the names of plants. I know, however, of only one plant in Sweden which has been given two names and that is the dog rose; one is for its flowering period and one for the time when the fruit is ripe. In summer, when fields and groves are dressed in their finest array, the dog rose is one of the most beautiful of all flowering plants. In autumn, when the fruits are bright scarlet, we call them hip shrubs, and their beauty is almost as great. Hips are interesting to study. The large fruit in the accompanying illustration belongs to the *Rosa rugosa* species, often planted as a hedge, the small round one, *Rosa pugeti*, and the one with the stem is *Rosa grisea*. The home gardener may notice some brownish-red hips on a cultivated bush. They hang like ox-blood red earthenware vases in long rows on bending branches.

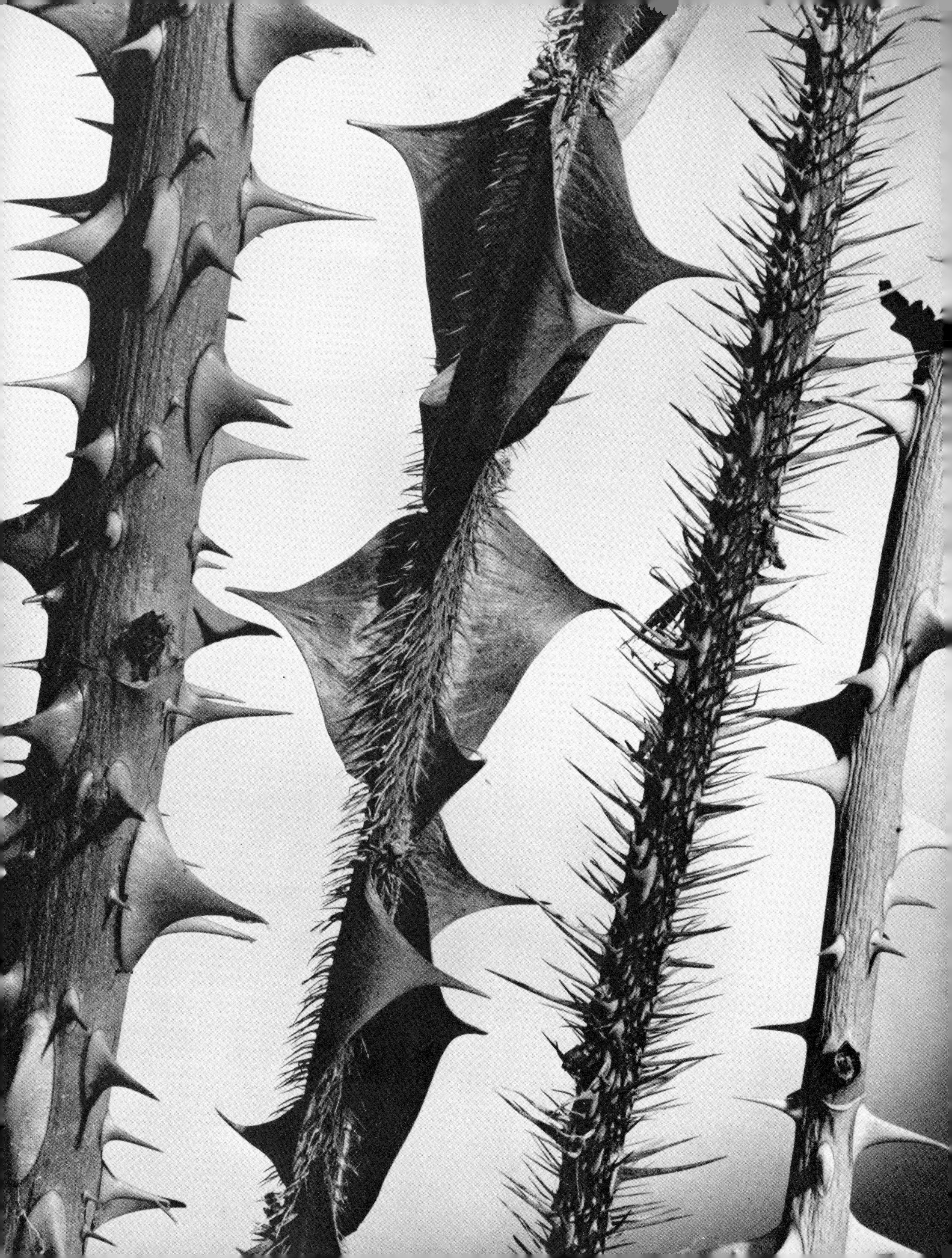

43

Some rose species have glandular filaments, and a hip from such a species is shown on page 56. It was taken from a bush in the park of an old manor house. The glandular hairs are red with white knobs, and in the slanting rays of the evening sunshine they give the hip a singular charm.

The flowers and the sharp thorns have given the dog rose its summer name of "thorn-rose." The thorns vary as much in form and appearance as the flowers and hips. A few species are shown opposite. The most splendid thorns with wide bases and asymmetrically tapering points belong to a species often cultivated, *Rosa omeiensis*. The thorns are crimson as are the young branches. The branch with the evil-looking fine thorns is from a species called, very appropriately, *Rosa spinosissima*, that to the far left is the *Rosa rugosa*, and that to the far right, *Rosa canina*.

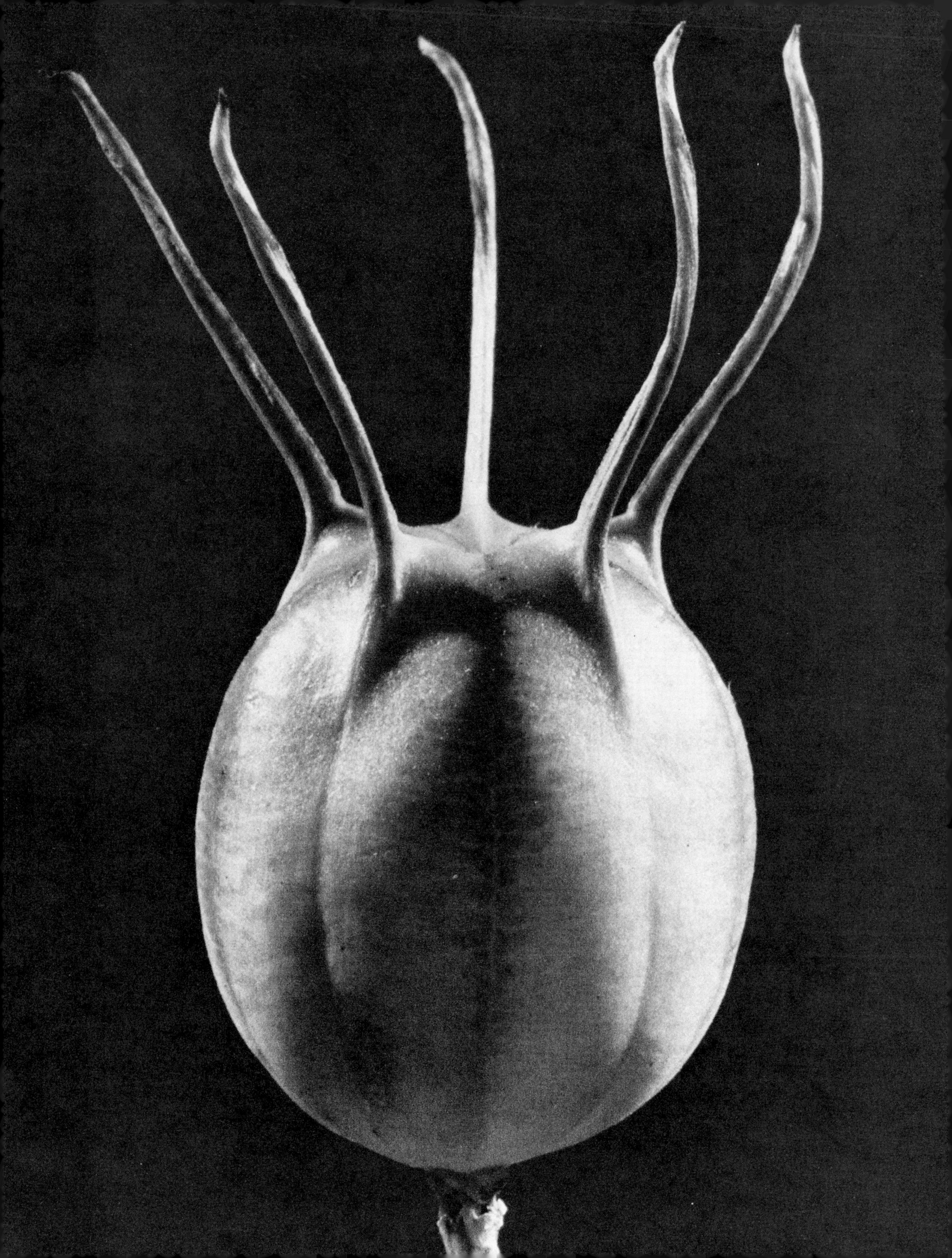

NIGELLA DAMASCENA

46 · NIGELLA HISPANICA

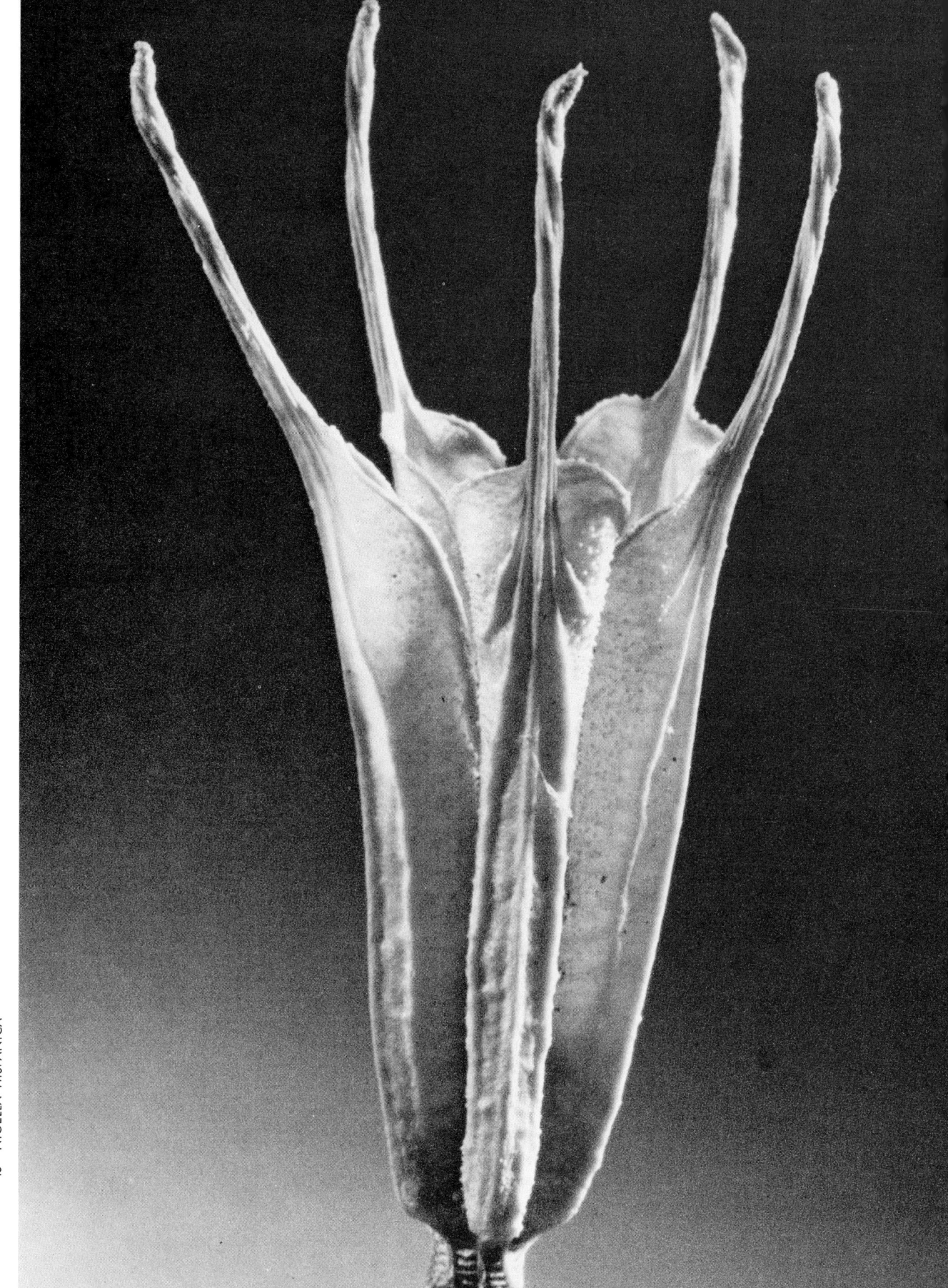

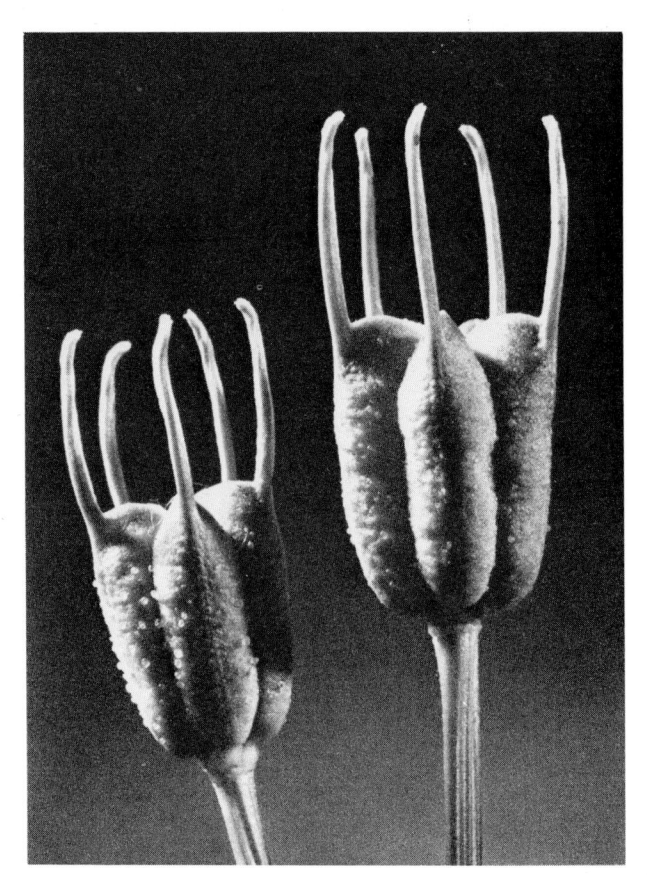

47 · NIGELLA SATIVA

*Nigella* species do not grow wild although some of them are cultivated as ornamental plants, especially the *Nigella damascena*. The English name, Love-in-a-mist, refers to the fruit, which is enclosed in a green network. Another English name is Devil-in-a-bush, inspired by the horned capsules peering from a bush of finely-divided involucre, while in France, it is called Cheveux de Venus. The blue-white flowers are beautiful, and the plant is worth growing for them alone, but the fruit is captivating with an inflated seedcase, the five "horns" as long and slender as spider's legs, and the involucre, which is first spread out under the flowers, and later wraps its protective lobes around the fruit.

Two other species of *Nigella*, from a botanical garden are included. One, *Nigella sativa*, has its five horns pointing stiffly upward, direct continuations of the outer edges of the five follicles. The other, *Nigella hispanica*, is more graceful, with twisted points on the long horns, which spread apart as the fruit ripens.

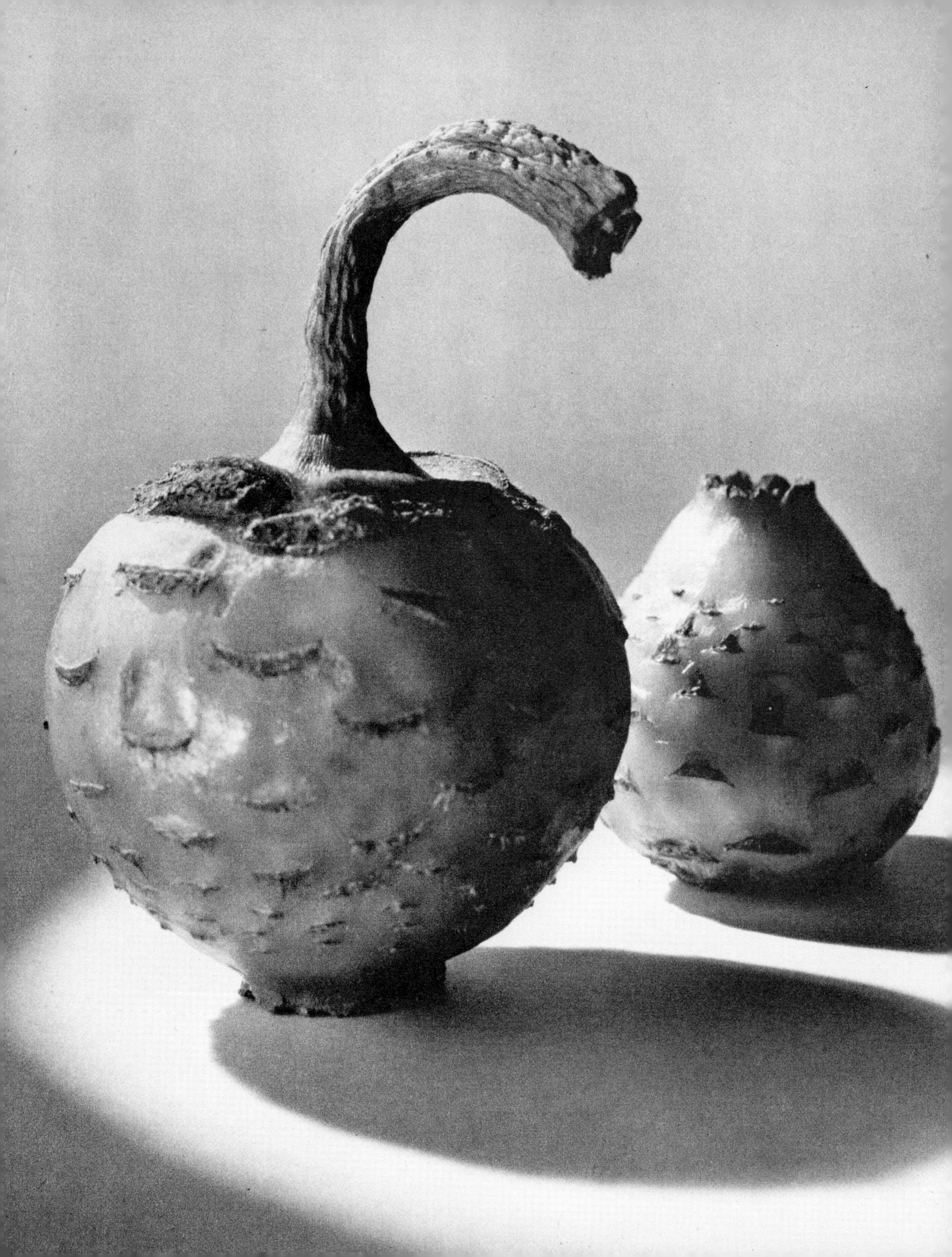

49 · NUPHAR LUTEUM

There was a pond with water-lilies in the garden of the house I lived in as a child, and for years I saw their white flowers and beautiful leaves floating on the water, and the brownish-red leaf buds, rolled up towards the middle, like a double-barreled gun. But I was sixty years of age before I discovered that the fruit was well worth examining. This is due to two characteristics of this plant, for soon after flowering, the stem bends and the fruit dives under the surface toward the bottom of the pond, and also because the withered petals, sepals and stamens hide the fruit for quite a long time. When they have fallen off, the scars they leave form a decorative pattern on the pale surface of the fruit. The fruits shown opposite belong to the *Nymphaea candida* species; they are almost egg-shape, the base being broader than the top. The other species of white water-lily, *N. alba*, has an almost spherical fruit, somewhat flattened, with a broad stigma.

The yellow water-lily *(Nuphar luteum)* cannot compare in beauty with the white species, but that it, too, is beautiful cannot be denied. The powerful pistil, in which the shape of the fruit may be recognized, dominates the center of the flower. The stigma is like a many-pointed star. Dense, concentric circles of stamens surround the pistil. As it grows in size and beauty, stamens and petals wither and fall. The five strong, convex sepals remain to support and protect the fruit, which is like a big-bellied flask or decanter, with a narrow neck. It terminates upward in a wide mouth. The surface is embellished with radial striae. It is like a dark

63

48 · NYMPHAEA CANDIDA

50

green stoneware vase, symmetrically designed, with occasionally, a slightly bent neck. Those who are interested in plant anatomy should study the inside of the fruit. It is constructed like an orange, with many sections full of pulpy mucilage, with bubbles of air around the small, smooth brown seeds. The bubbles keep the sections afloat and help to carry the seeds away on currents of water.

It is not surprising that such beautiful plants as water-lilies have many names. The scientific names, *Nymphaea candida* and *N. alba*, which, in English, mean the white water nymph, originate from Greek mythology. A fourteenth or early fifteenth century English herbal calls *Nuphar luteum* "water folys fot," because the leaves, it is said, are similar to the "medyl coltys fot." A more modern English name is brandy-bottle.

51 · BLUMENBACHIA SPIRALIS

*Blumenbachia spiralis* should be handled with care. Its sting is almost as painful as that of our stinging nettle. The fruit, which is slightly smaller than a walnut, is similar to a twisted melon. While growing, they are pale green and have a meat-like consistency, but when ripe they are grayish-white and dry, they split along the spirals like orange slices, and release their small, dark seeds. *Blumenbachia* is a native of South America, but the fruit shown here grows in a botanical garden as a luxurious carpet, eight to ten inches high.

53

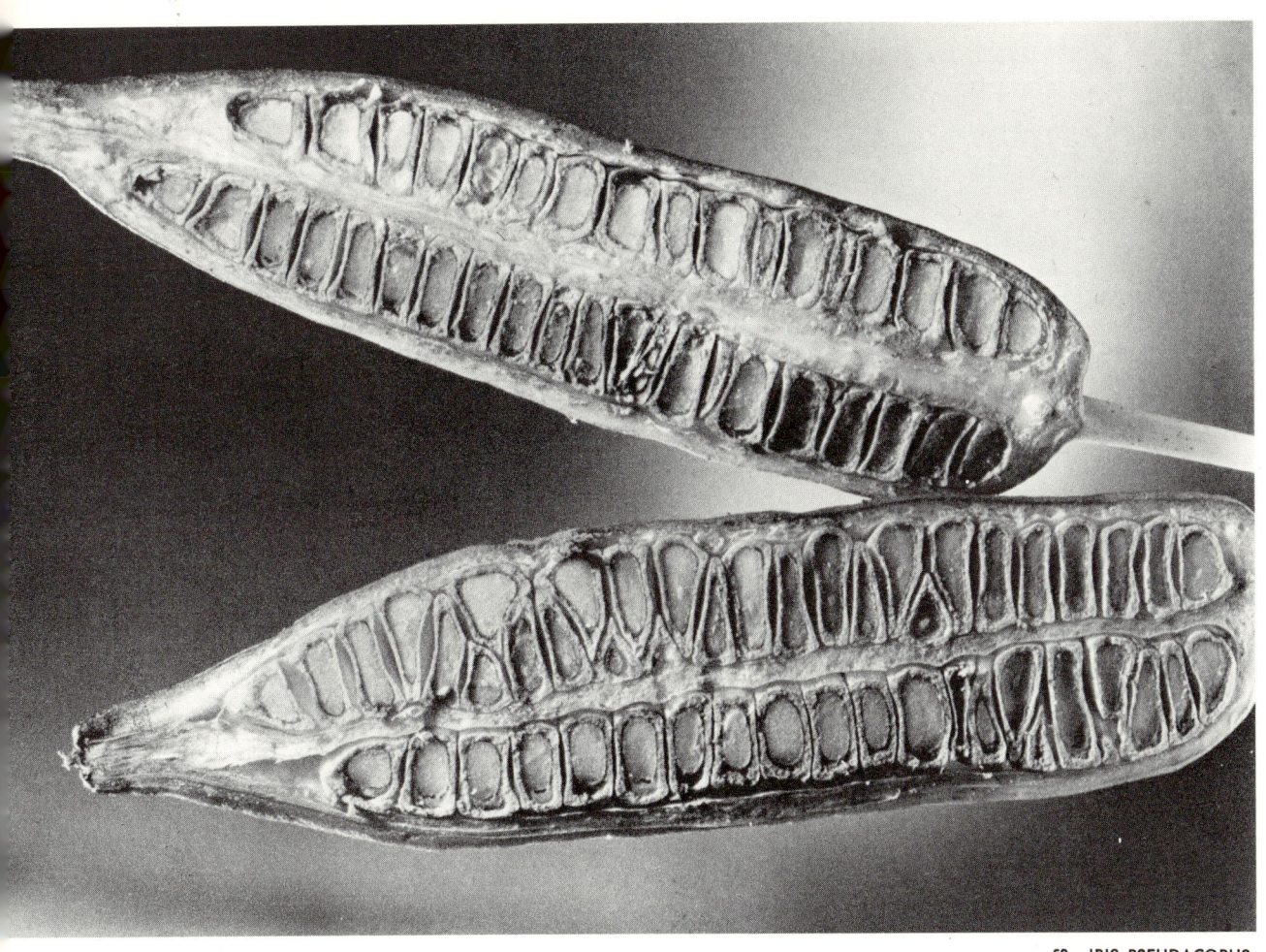

52 · IRIS PSEUDACORUS

The yellow flag (*Iris pseudacorus*) bears fruit profusely, large bunches of cucumber-like capsules, often so heavy that the slender stems cannot uphold their weight, and bend lower and lower toward the ground as the seeds ripen. When the seedpod has burst at the top and the pointed lobes begin to turn backward, the flat, glossy brown seeds can be seen, packed on top of each other, contrasting sharply with the white, silky inner side of the walls. A half-ripe fruit, cut along the middle, shows how crowded the seeds are. Each pod contains three rows of seeds.

54 · ESCHSCHOLTZIA CALIFORNICA

The sepals of *Eschscholtzia californica* are fused together, and when the time approaches for the flowers to open, the sepals burst open and hang like a funnel on the point of the flower until they fall off. At dusk, the yellow flowers close so tightly that they again look like buds. The name of the flower means "sleepy-head," but whether that is due to the way the flower closes at dusk, or because the sepals look like a night-cap, I am not certain. Another name is "candle-snuffer," which probably refers to the similarity between the sepals and the long-handled metal funnel formerly used to extinguish home-made candles.

The photograph above shows a bud still in its cap. The other three specimens show the fruit in various stages of development. The little fringes below the bud and fruit are pale red, which harmonizes well with both the deep yellow flower and the pale green fruit. Mature fruit bends in a slight arch and splits longitudinally when the seeds are ripe.

55 . PARNASSIA PALUSTRIS

The connection between Parnassus and the scientific and English names of *Parnassia palustris* is not clear. Perhaps it grows on the slopes of Parnassus, or perhaps the ancient Greeks who, at the dawn of history, gave the mountain range to Apollo and nine goddesses of song as a dwelling-place, considered this flower an appropriate tribute to the god and his joy-giving muses. In any case, it is an unusually charming flower; one might call it the anemone of late summer. It is remarkable in two other respects. Before the stamens are fully grown, they bend inward and touch the stigma of the pistil with their anthers, a shy kiss of youth on the flower maiden's lips. Further, there is a curious formation between the five stamens, a small disc or scale, the upper edge bordered with thin fibers with knobs at the tips. Botanists call these staminodes, and their function is to produce honey, or nectar to use the terminology of Parnassus' gods. They can be seen clearly in the illustration, but must be seen in reality before their full beauty can be appreciated. The strong pistil grows rapidly after it is fertilized, and toward the end of the flowering period it dominates the flower. In shape it is something between an egg and a cone. When the fruit is ripe, it opens at the top and the walls become wrinkled laterally.

57 · ACER PSEUDOPLATANUS

58 · ACER PLATANOIDES

The golden clusters of Norway maple, flower on bare branches early in spring, and the flaming red leaves of autumn are among the joys of adults, while the fruit—maple noses—are fun for the children. But grown-ups can also find pleasure in the beautiful fruit. In autumn, when they have attained their full size, the heavy clusters of fruit can easily be seen among the leaves, but while they are growing they remain hidden. They are perfectly formed even at a very early stage. The fruit shown above is unripe, less than an inch long. The pistil with its divided stigma—a two-headed eagle on outstretched wings—is still in its place. I have read somewhere that Helen of Troy, King Menelaus' wife, was dressed in clusters of maple flowers at her wedding. There is no doubt that she, "the most beautiful woman in the world," chose the prettiest things she could find to wear.

The white spots on the fruit are short glandular hairs. The veins, which cannot be seen very clearly in young specimens, gradually become more distinct. When the clusters, which often hang on the tree throughout the winter, have been exposed to the weather and become more transparent, they may have the same delicate network appearance as an insect's wing. The picture opposite is of such a fruit, the sycamore, *Acer pseudoplatanus*.

59 · LONICERA SYMPHORICARPUS

*Lonicera symphoricarpus* is cultivated for the sake of its white fruit. The flowers, which are small and insignificant, are clustered together in a small, pale red raceme. Each little flower looks like a miniature goblet. A fully-grown ripe fruit is a white berry with a glossy surface, but it is not often that all the berries in a cluster grow to full size; many stop growing. It is interesting to study all the transitional forms from the flower goblets to the large berries. They are like a series of white vases of different size, standing in rows along the stem of the cluster. The shrub is planted in hedges or alone. The specimens photographed here are from the gardens of Tullgarn Palace, where the bushes grow in a large mass of shrubs. When the leaves have fallen in the autumn, the entire shrub is dotted in white against the dark branches.

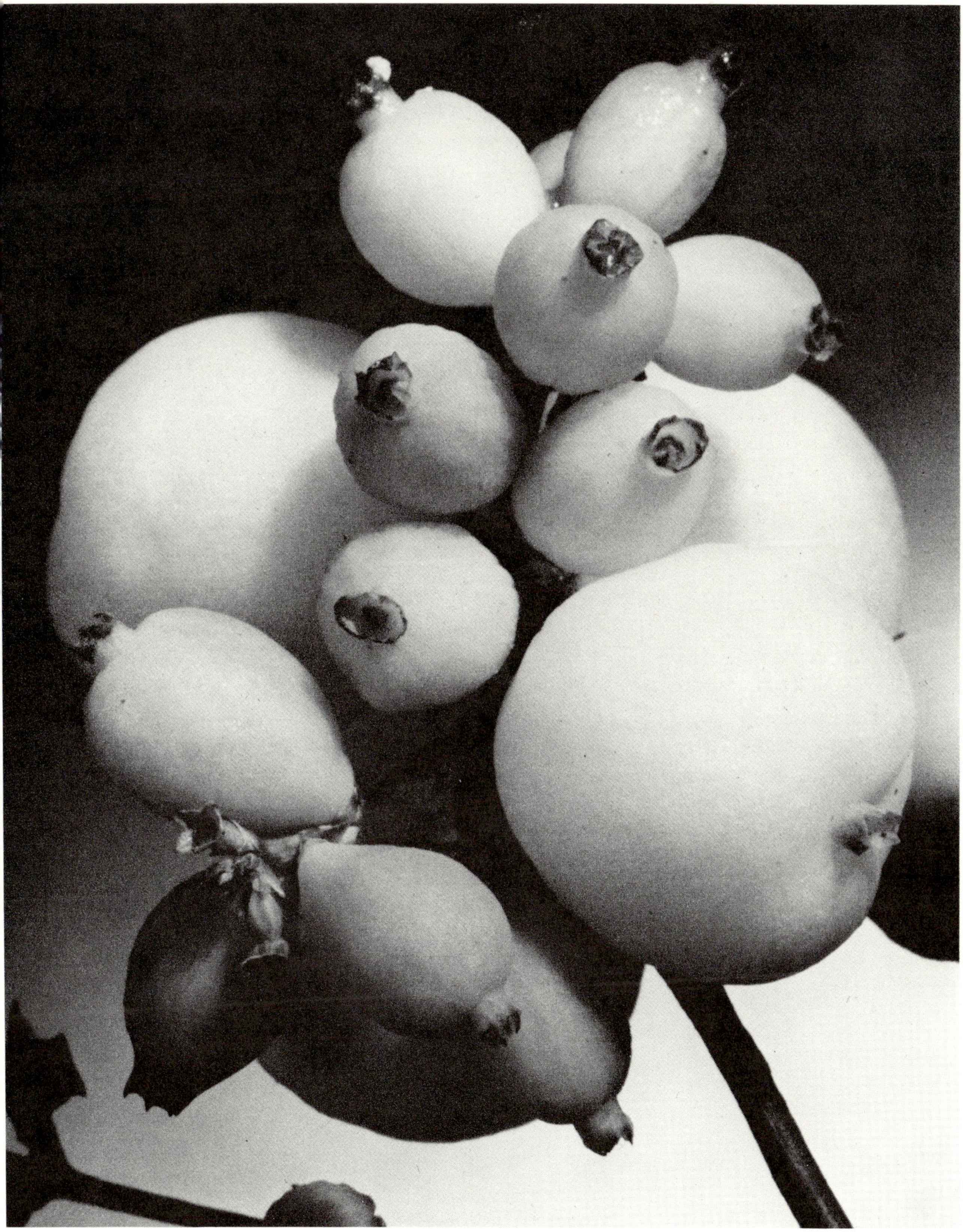

60

Plants belonging to the parsley family, Umbelliferae, are very common. Their beauty lies in their stately growth and the mass effect of their numerous small flowers. Wild chervil is the most common species, and when it adorns the hedgerows with its white flowers, we know that summer is really here. The fruit is the same as that of all other members of the family; it is a twin-seed, and the seeds are joined together longitudinally along a common dividing wall. When ripe, the seeds first loosen from the bottom and hang for a time on fine threads which arise when the dividing wall or the beak, as it is called, splits at the top, until at last they are dispersed by the wind. Wild chervil seeds are blackish-brown, glossy and fusiform, and when studied under a magnifying glass they are found to have a fine pattern of spots very close together, between wavy lines. The double-horned tops—the remains of the stigmas—make the twin fruit look something like double-necked glass decanters.

It is fascinating to study the variations in the structure of the fruit of the Umbelliferae. Many are longitudinally striated with more or less prominent ridges, others are winged, those of the carrot, spinous; some are elongated, others barrel-shape. And the ever-puzzling question arises; why does Nature indulge in all these variations?

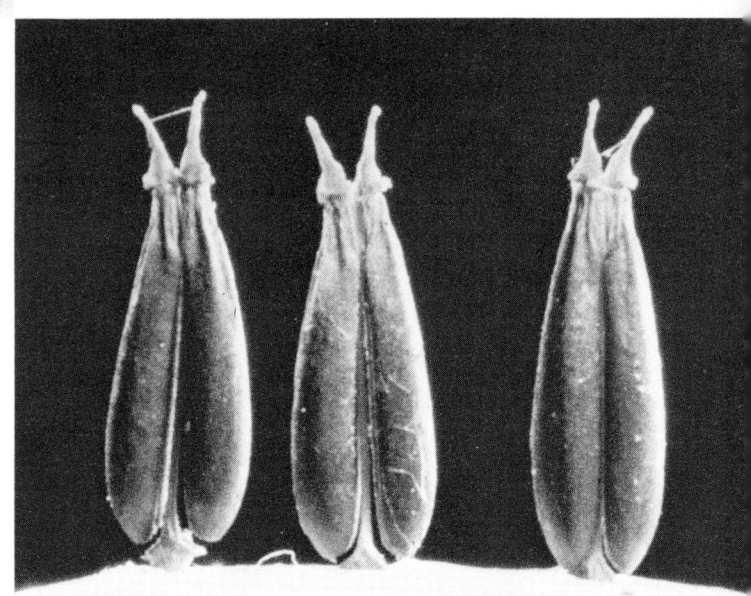

61 · ANTHRISCUS SILVESTRIS

63 · FILIPENDULA ULMARIA

*Filipendula ulmaria*, known commonly in English as the meadowsweet, was once known as meadwort and bridewort. It is a stately plant with luxurious foliage and white flowers in large panicles. Gerard's *Herball* also lists Queen of the Meadows, which is the name given the plant in France and Germany (La Reine des près and Die Wiesenkönigin). An old gardener I knew in my childhood used to rub empty behives with meadowsweet to attract the queen bee so that swarming bees would not fly away. Gerard says: "the floures boiled in wine and drunke, doe make the heart merrie," and "the leaves and floures farre excell all other strowing herbes."

The fruit is in dense clusters of small balls. The crooked and twisted carpels, each with three or four seeds, give the ball a spirally striated appearance. They remind one of long-fingered hands tightly clasping each other.

64 · SPARGANIUM SIMPLEX

The unbranched bur-reed *(Sparganium simplex)* called hedgehog bud, grows in damp meadows by lakes, at times directly in the water, and is soft and porous, like an aquatic plant. The male and female flowers are separate from each other, the males in a long row at the top of the stem. In the bud stage, the males are rather dark in color, but when the flowers are open the many stamens give them a pale green lustre. Eventually, only some insignificant withered remains are left, as in many species of *Carex.* At the same time, the heads of female flowers flourish until they have the droll appearance shown in the picture above, resembling a flock of cackling leghorn hens.

76

65 · NICANDRA PHYSALOIDES

*Nicandra physaloides* is called Apple-of-Peru. It has rather large blue flowers, which are beautiful in themselves, but my interest was aroused by the sepals. They are a curious blackish-green color, and similar in shape to an arrowhead with raised edges. If the five sepals are held apart, the fruit becomes visible. It is a flattened sphere about as large as a cherry, with a surface which is faintly bulbous, owing to the pressure of the seeds inside. The sepals grow simultaneously with the fruit, and the color becomes paler, first a yellowish-brown and later, silver, while the network of veins becomes clearer. One of the fruits in the illustration has burst open and its black seeds have begun to run out.

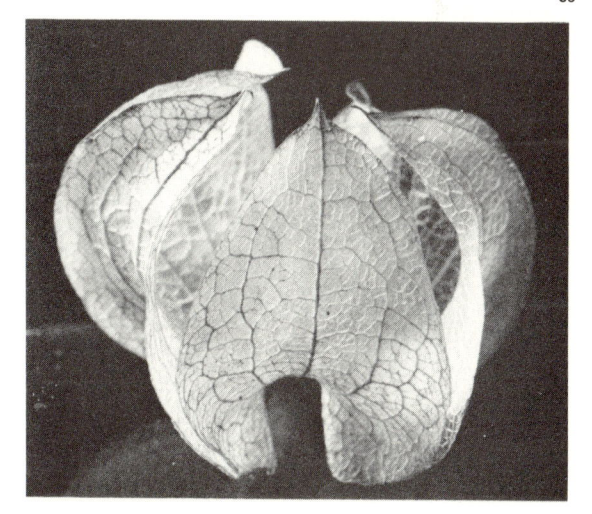

66

77

67 · HIBISCUS ESCULENTUS

68 · HIBISCUS TRIONUM

The *Hibiscus* is rich in species, but only two of them grow wild in Europe. Most species are tropical plants, a few, however, are grown in pots in temperate countries. The best known is the Chinese rose, *Hibiscus rosa-sinensis*. It flowers profusely, with large red blooms, which soon wither and I have never seen one bear fruit. In the tropics, the flowers are used in cosmetics; women stain their eyebrows and the edges of their eyelids black with the juice of the crushed flowers. The fruit of *Hibiscus trionum*, illustrated above, grows wild in southern Europe and in Africa. The sepals are double; immediately below the flower is a wreath of the ordinary type, while the upper sepals grow together, inflated into a small balloon. Dark ribs enclose the balloon. They are in sharp contrast to the pale lustre, and to my mind they underline the beauty of the plant in the same way as the black juice of the Chinese rose emphasizes the beauty of women in the tropics.

This plant, which belongs to the Mallow family, is sometimes called *Hibiscus* and at others *Abelmoschus*, but in both cases it has the species name *esculentus*, meaning edible. It is an annual plant, native to the East Indies. The young, unripe pods are called *okra* or *gumbo*, and are considered a great luxury. When the fruit is ripe, the pod becomes hard and woody. The seeds are used as a substitute for coffee, and medicinal decoctions are made from the leaves and roots. Owing to the many uses of the plant, it is eagerly cultivated. The fruit is also beautiful, with its double, pale, almost white stripes or ridges, on a dark background.

At an early stage, the fruit is pale brown. The white stripes appear when the fruit begins to open; they are the edges of the seedcases, which burst open and turn their white surfaces outward. In the half open fruit, long rows of almost spherical black seeds, about as big as peppercorns, can be seen. Shake the fruit and it sounds like a rattle. The cracks widen gradually and eventually the seeds fall out.

69 · LONICERA CAPRIFOLIUM

The banks of white honeysuckle on the gables or around the verandahs of houses and cottages at dusk on a summer's day create an extraordinary atmosphere. The scent is almost overpowering, and heavy moths fly from flower to flower, thrusting their tongues deep into the nectar at the bottom of the spurs. But it is not only at flowering time that honeysuckle should be admired; follow the plant's growth from spring to autumn and you are sure to make many interesting discoveries. The stems grow in a curious manner as the leaves, opposite each other on the stems, are joined together to form a shallow dish. As the stem continues to grow, it thrusts its way through the dish and forms a new leaf bud as shown in the picture below. The new bud is nearly triangular and has sharp edges. After a time it opens into a new dish, at right angles to the previous one. The flower buds growing in a ring around the stem in one of the dishes, seem to be supporting the next dish. When the flowers have withered and fallen, the tiny green ovaries can be seen. They grow slowly and do not mature into red berries until the end of summer when they finally have grown as large as big cowberries, appetizingly arranged on the green serving dish. It is not often that nature creates anything as beautiful or more perfect in design.

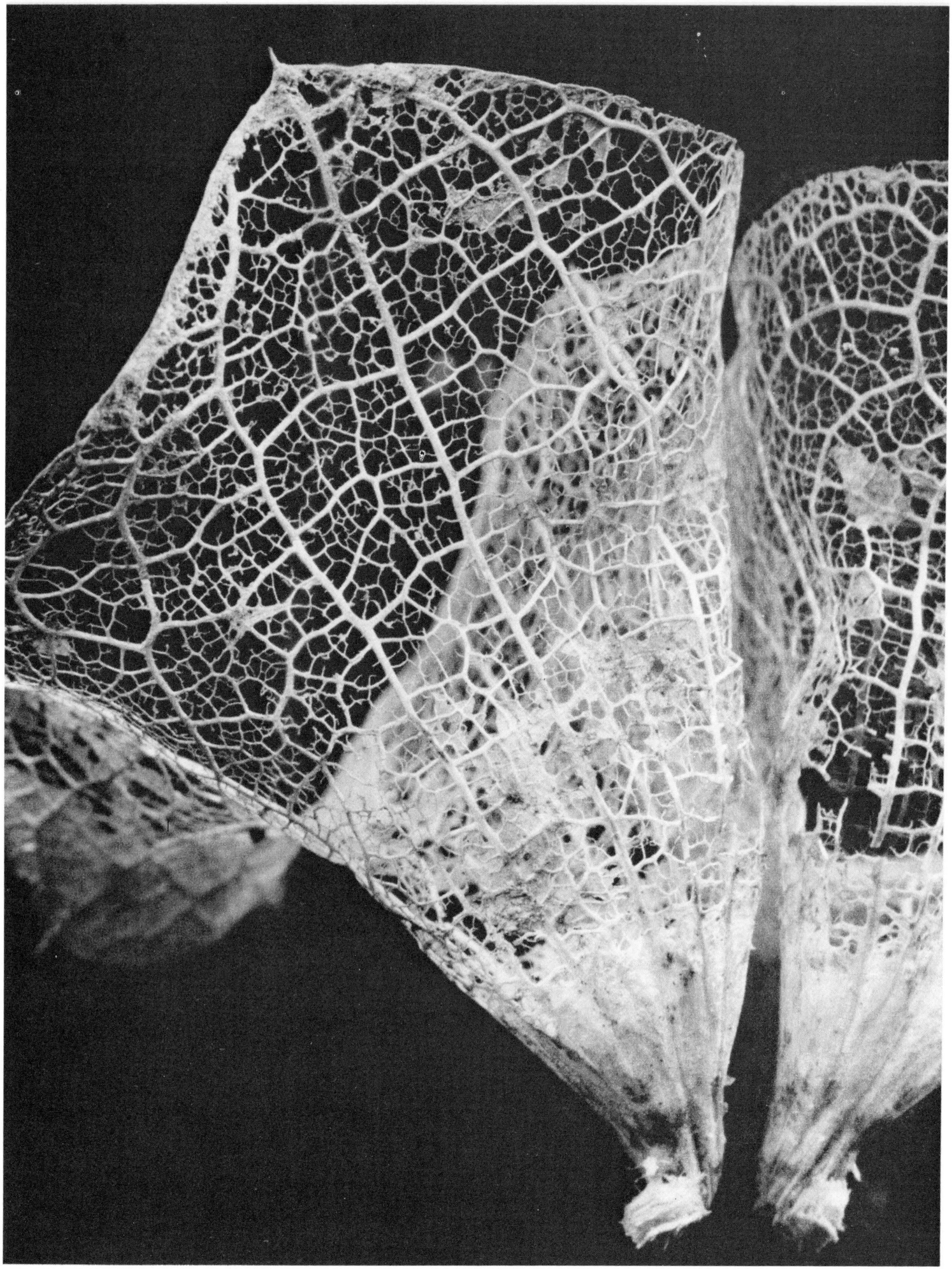

82

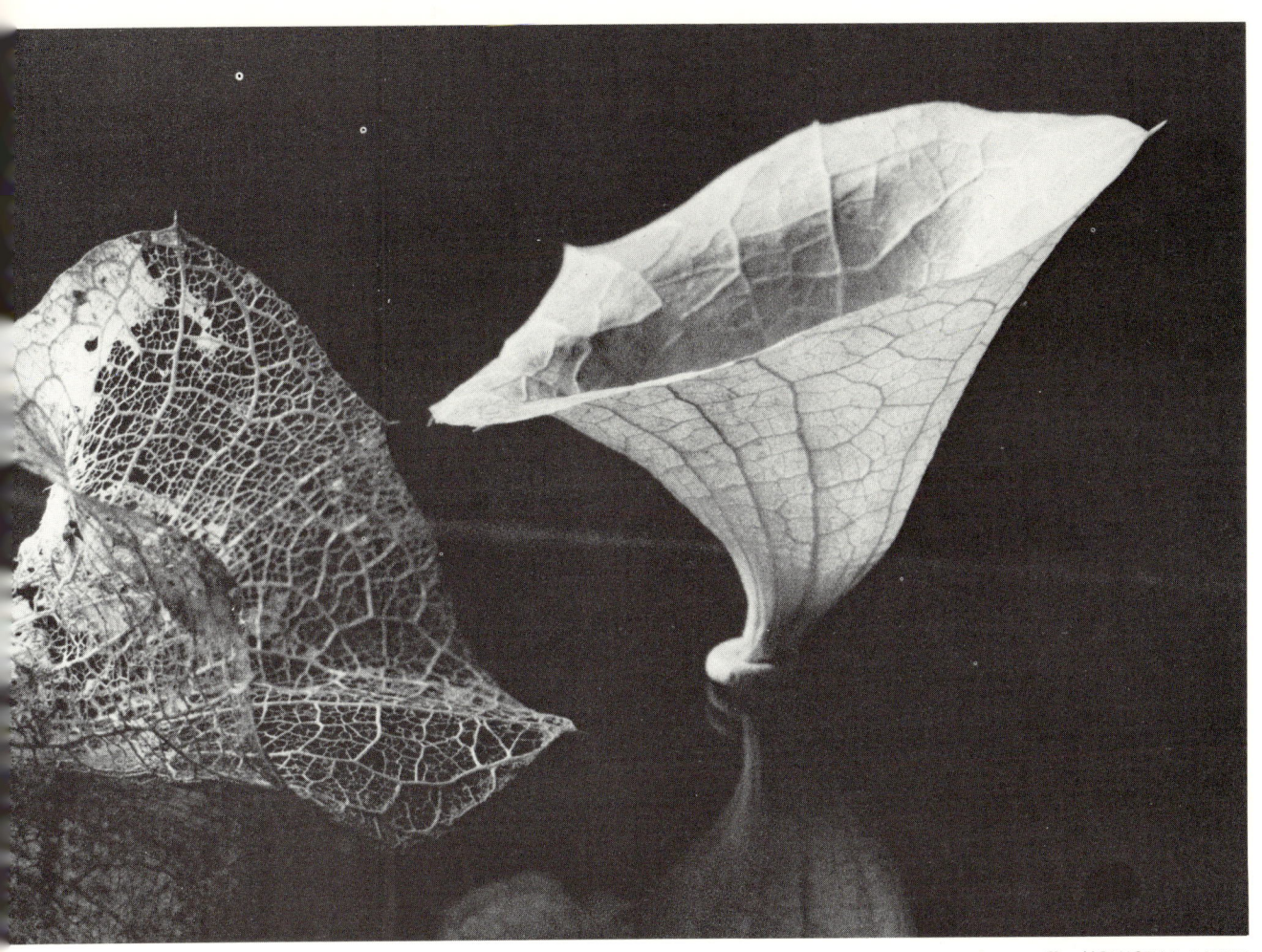

71 · MOLUCELLA LAEVIS

*Molucella laevis* takes its name, Molucca-balm, from its native land, Molucca—also called Spice Island—in Indonesia. It can be cultivated as was the specimen shown here. The flowers, which are similar in form to those of the white deadnettle, grow in five or six rings around the stem, and there are a large number of such rings on the long stems. There is nothing remarkable about the flowers, but the goblet-like calyx is, therefore, all the more interesting. Trumpet-shape like an old-fashioned phonograph horn, their wide openings are turned toward all points of the compass. In the illustration above, a calyx is standing like a goblet with the opening upward. It has a complicated network of strongly marked veins, which become more distinct as the fine tissues between the veins decay and disappear.

73 · MALOPE TRIFIDA

*Malope trifida* is a plant which grows wild in the Mediterranean countries, but which is often cultivated elsewhere. It is closely related to our common mallows. The pink or white flowers are large and showy. When the petals have fallen off, the double wreaths of sepals close at the top and thus protect the seeds. The plant is dull brown, but if the sepals were separated, the splendid fruit shown above could be seen within. The beautifully striated seeds are crowded closely together on their central pillar.

74 · SEED VESSELS

Sepals are usually one of the less striking details of a plant; with respect to both color and shape they seldom bear comparison with the petals. But when the sepals are joined together, gamosepalous in botanical terminology, they often form complicated goblet-shape formations of great interest. The variations are often very important in the classification of different species, but the calyxes are worth examining for their own sake. Above is a picture of such goblets at the stage when the corolla has just withered and fallen. The small seeds at the bottom of the vessels are well protected. The goblet on the far left belongs to the common hemp-nettle, *Galeopsis tetrahit*, and the next, to a plant from Syria, *Phlomis viscosa*. The other two are from a very thorny plant with leaves similar to those of the acanthus, *Morina longifolia*, from the Himalayas.

85

75 . CALLA PALUSTRIS

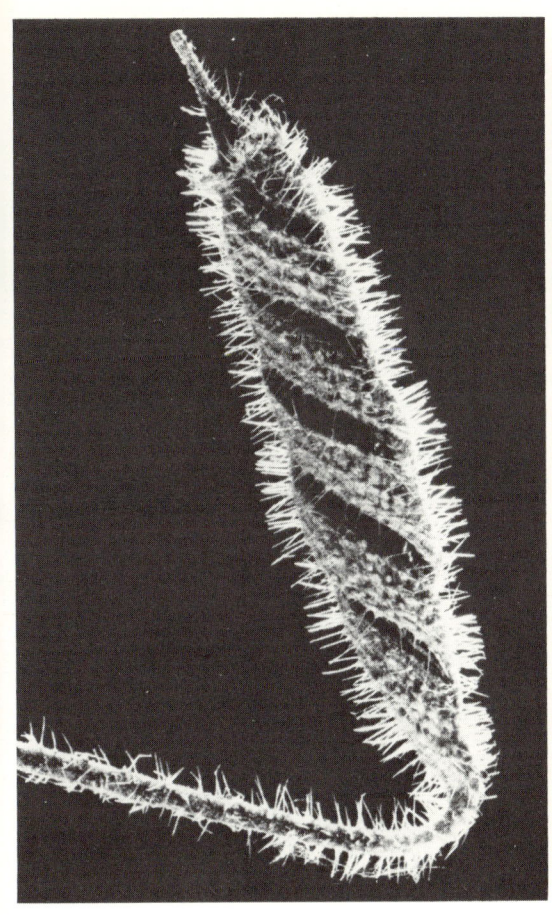

76 · CAIOPHORA HIBISCIFOLIA

77 · CAIOPHORA CERNUA

I am always glad to find a bog in the woods with a dense carpet of *Calla palustris*. The large, glossy, bright green leaves and the juicy creeping roots give the impression of seething life. If it happens to be the height of summer, when the plants are in flower, my joy is also heightened. The greenish-white club (spadix) at the top of the stem is seen against a dazzlingly white background (spathe). It looks like a candle in a sconce. A few weeks later the scene has changed. The flowers have set fruit, the spadices have swollen and changed color, and are now bright red, while the white spathes have turned green. The general picture is the mosaic pattern of the fruit stage. The spadix might be likened to a cluster of naturalistic human breasts drawn by a surrealist.

The botanist who gave the *Caiophora* genus its scientific name was clearly thinking of the stinging hairs of the plant. The Greek word *caio* means "I burn" and *phoros* means "bringing." But he might also have had the fruit in mind when he named the plant, because of their uncommonly characteristic appearance. They are three to four inches long when full-grown, and twisted into a spiral. When the time arrives for the seeds to be dispersed, the spiral loosens and fissures appear through which the seeds escape. Some species from Argentina are climbing plants, which bloom abundantly and beautifully, and ripen even in this northern climate unless frost comes too early. I chose two species for my photographs; the one with the dense stinging hair is *Caiophora hibiscifolia*, the other *C. cernua*.

87

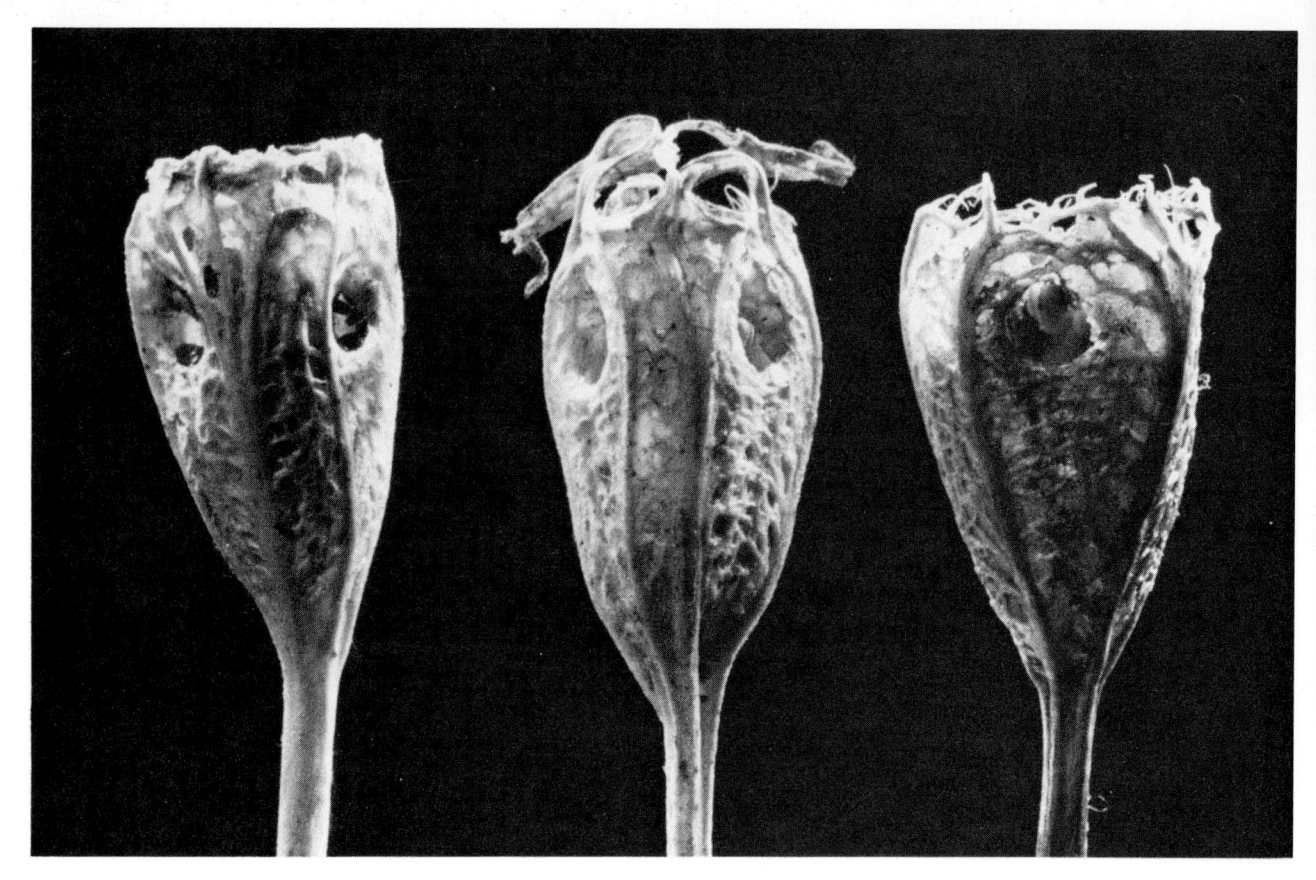

78 · CAMPANULA PERSICIFOLIA

The fruits of the bellflower stand upright on their stems, or they may droop; each species has its own type. When the seeds are ripe, three small holes appear in the walls of the seedcase to prepare a way for the seeds to escape into the world. These holes are so located that the seeds will not fall close to the parent plant, but will be thrown some distance when autumn winds shake the stems. The holes in the drooping capsules are near the base while those in the upright capsules are near the top. Disintegration sets in when all the seeds are dispersed and the fruit has served its purpose. The network of nerves becomes more distinct, the thinner parts crumble away. The veins are most resistant, and at a certain stage only an open basket remains in some species. The baskets or cages shown opposite belong to *Campanula carpatica*, often grown as a border plant. The photograph above is of the seed capsules of the large bellflower *(Campanula persicifolia)*. The surface is divided by longitudinal veins, between which is a fine network in relief. The edges of the holes are smooth, so that the small seeds can easily slip out. A few seeds can be seen in the opening in the fruit at the left.

79 · CAMPANULA CARPATHICA

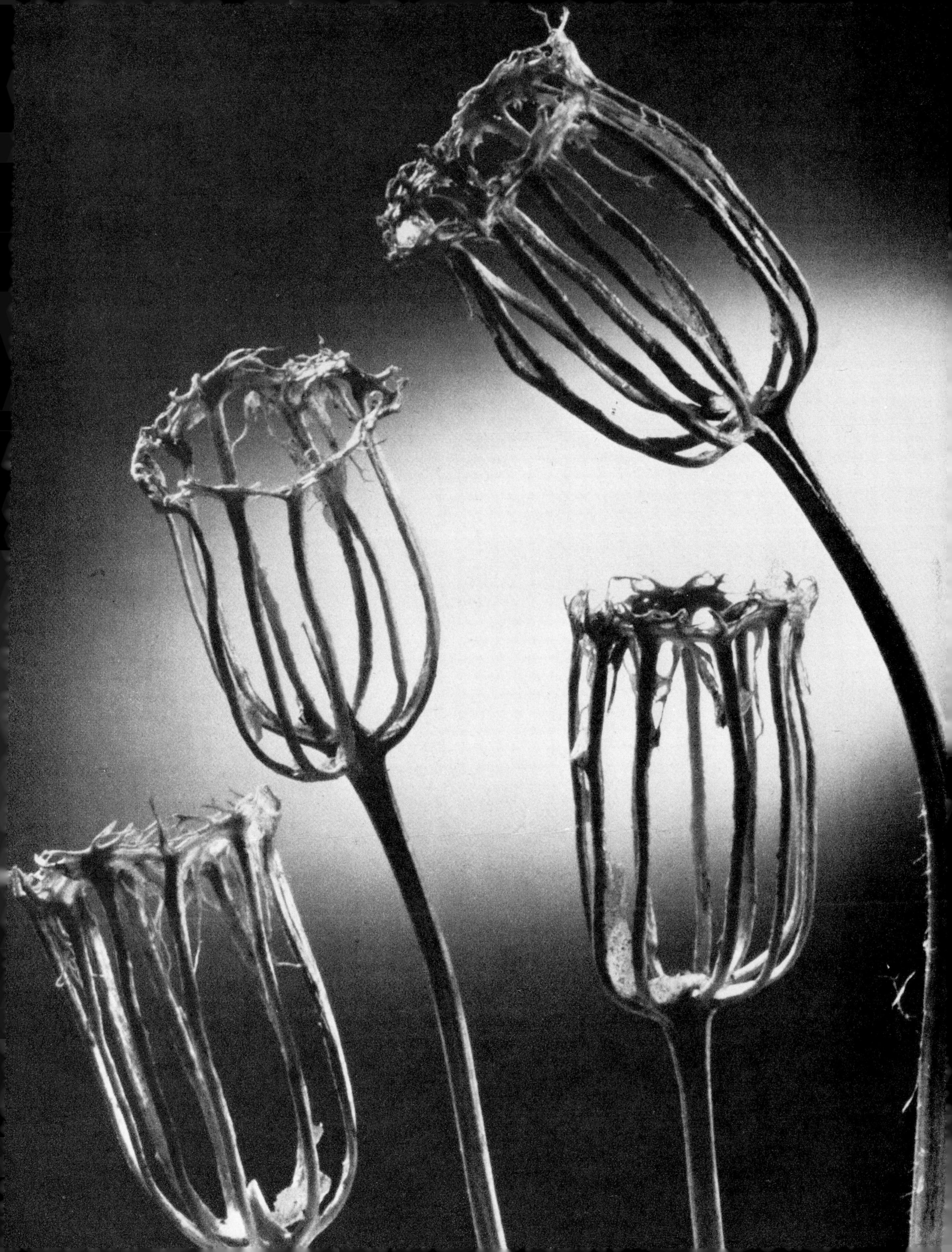

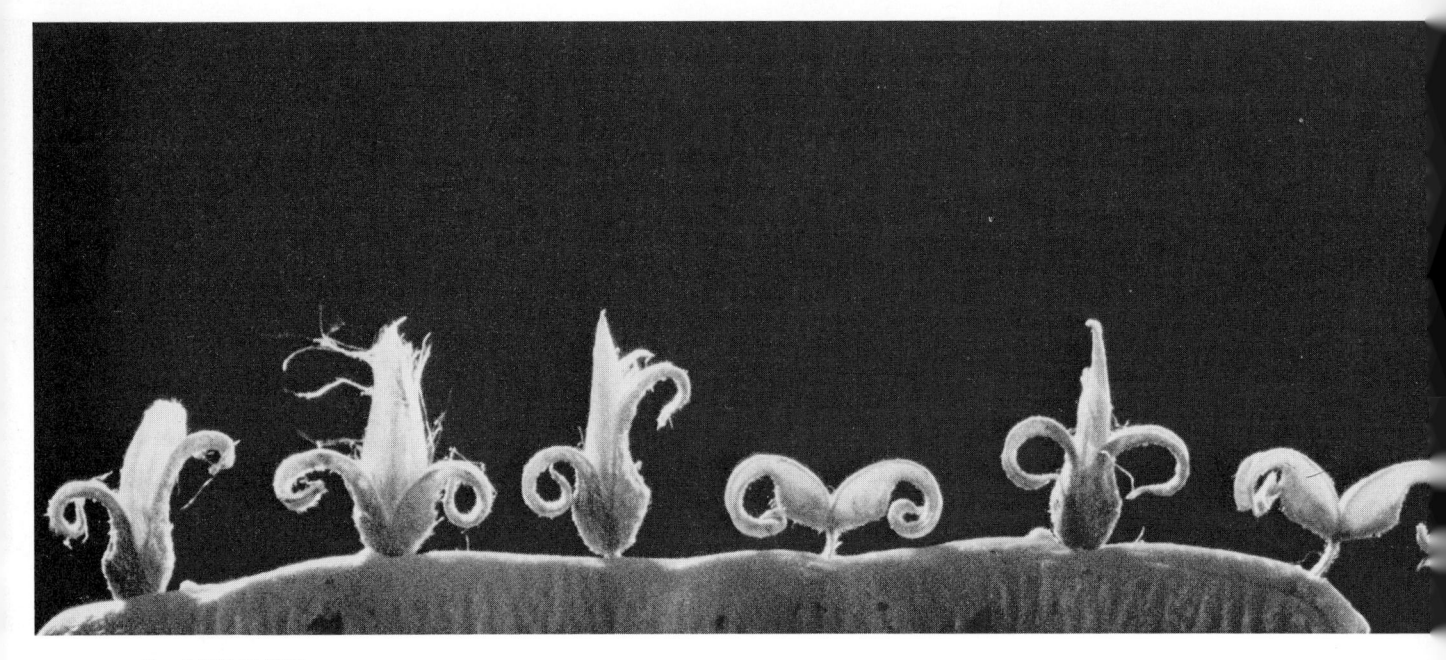

80 · SALIX CAPREA

81

82

83

The sallow shrub which has the misfortune of growing near a town, will be the victim of its own beauty. The twigs are usually gathered so early that the black scales on the buds must be removed so the attractive white silky hair may be seen. Waiting for the extremely beautiful fruit, however, would be far more rewarding and these could simply be cut from the twigs, leaving them on the shrub unharmed. The female catkin consists of a large number of small flowers, each producing numerous seeds. The withered flowers in the illustration above with their swollen bases where the seeds develop, look like a flock of cackling geese (Illustration 83). The ripening seeds, exerting pressure from the inside, burst at the top. From a tiny white point the rupture continues downward until the fruit opens wide in two graceful arches, with a white bush of fine hair between them. The seeds are then detached and fly away on their silky filaments and only the empty, arched seedcases remain leaving a general impression of untidiness. If a single catkin is studied, one finds a jumble of empty seedcases. One of the illustrations shows a series of separate flowers in different stages of development.

84 · TARAXACUM VULGARE

85   86 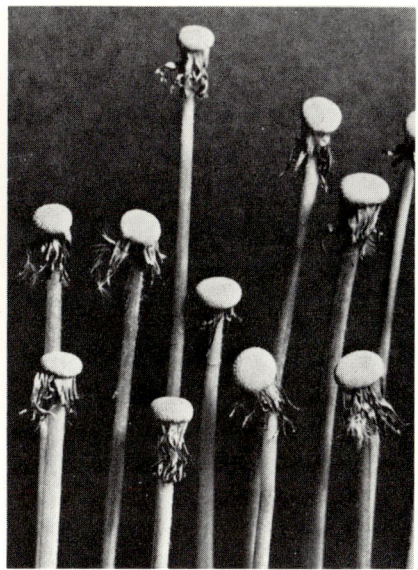  87

The dandelion is incredibly vital; if the tiniest piece of root is left in the ground, a new plant will soon appear, remindful of the many-headed Hydra of Lerna—cut off one head and another appears! Many flowers can grow from a single root, and each flower can produce two or three hundred seeds. How many can an entire field of dandelions produce? No wonder the dandelion is the most despised weed in the garden. But let us forget that the dandelion is a weed, and regard it as a rare and coveted plant. It is well worth it. What can compare with a field of dandelions in flower, the bright yellow flowers—"Shock-headed dandelions, that drank fire of the Sun," as Bridges wrote—against a dark green background of leaves? The dandelion is well worth studying in detail. The robust, striated buds are surrounded by a lobed collar, the epicalyx, bending backward toward the stem. Just before the flowers are ready to open, the bud changes shape, from near spherical to conical. When the sun rises, the flowers open, when it sets, the epicalyx closes over the glorious yellow petals like a green nightdress, and this continues day after day until the flowers wither. The epicalyx then wraps itself tightly around the withered petals. The yellow remains of the flower are gradually pressed out and hang at last as a spirally twisted lump on the tip. The seeds can then grow and ripen, well protected. One of the illustrations shows a section of a flower with the seeds and their folded parachutes on long stems. The epicalyx begins to open from the top, and gradually, the parachutes open wide to form the unpopular, but wonderfully beautiful "fairy clocks." Examine the pale brown, striated seeds under a magnifying glass before they are borne away by the wind on their ingeniously constructed parachutes. When all the seeds have flown away, there remains a shiny white dome with numerous dark spots, where the seeds had been attached. The stems are often pale pink. Children amuse themselves by making necklaces and bracelets of the hollow stems. Their parents may also find it amusing to put a few stems in water and watch them split and roll up in curious patterns.

88 . ASCLEPIAS SYRIACA

90 . CYNANCHUM VINCETOXICUM

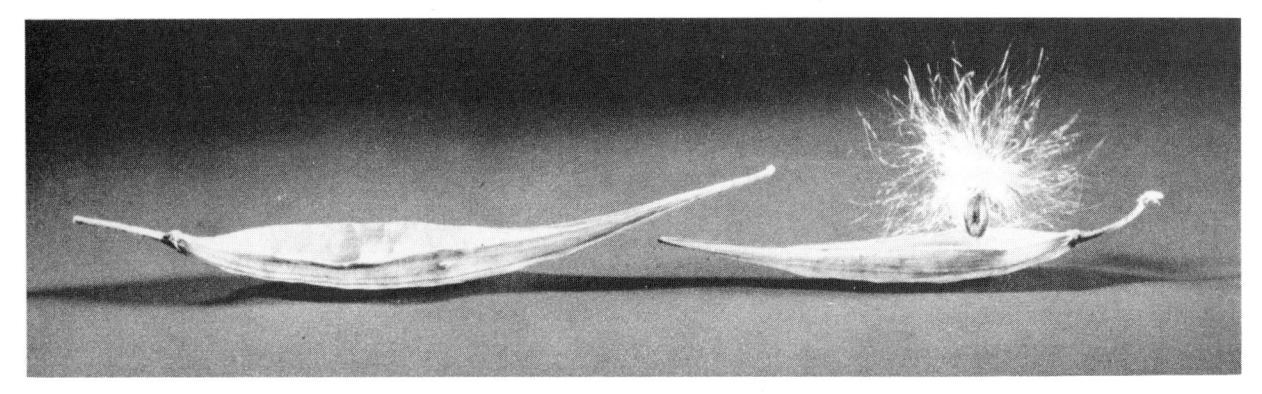

*Asclepias syriaca*, of the milkweed family whose generic name is derived from that of Asklepios the god of medicine, grows wild in North America. Its roots have been used in folk medicine as a cure for angina pectoris. Species of *Asclepias* are often cultivated for the sake of their beautiful flowers, and also as a textile plant. The fruit, with its curiously curved stem by which it is attached to the main stem, is shown on the opposite page. The fruit is full of seeds, lying along an elongated spadix, something like a spruce cone. In the photograph above, the *Asclepias* spadices have opened and the seeds are beginning to emerge. Attached to the brown seed is a silky tuft of hair, called Asclepias wool, which is collected and spun to make thread. The plant is also called the silkwort. There is yet another plant, *Cynanchum vincetoxicum*, closely allied to this species. The photograph below shows how the seeds, with their white down, emerge when the fruit splits into two halves. To the left is an empty shell, an elegant dish tapering from a broad base to a long point. The inside is glossy and white, the outside, pale yellow.

95

91 · TRAGOPOGON PRATENSIS

Of all the plants whose seeds sail away on the wind with the help of parachutes, *Tragopogon pratensis* (goat's-beard) takes the prize. The ball of downy brushes is larger and more airy than the dandelion's; it is so graceful, so elegant, that one cannot but wonder at this masterpiece of nature. The parachutes are funnel-shape, like umbrellas turned inside-out by the wind, which gives the head a geometric form. The ribs stick out slightly beyond the fine-veined umbrella. The elongated seeds are longitudinally striated, with tiny barbs on the ridges. The purpose of these barbs is to hold the seeds in place in the soil when they have landed. Goat's-beard flowers are also beautiful with their yellow petals, against which the stamens and their dark anthers form a striped pattern, but the flowers open only in the morning, hence their popular name, "Jack go to bed at noon." The buds and closed flowers are shapely, longitudinally striated cones, and the leaves, which are stem-clasping, form long, tapering spouts.

*Scabiosa stellata* is a relative of the scabious family, named from the Latin, scabies (itch), as the plant had been used medicinally for this disease of the skin. The flowers are assembled in a head at the top of a tall stem. The specimens illustrated here are from Portugal, and are in the fruiting stage. The structure of the flower is most curious. It has an inner and an outer calyx, and after the plant has flowered, the latter becomes a very beautiful bowl, as thin as gauze. It is kept open by about forty fine nerves, much like the ribs of an umbrella. A five-pointed star stands like a silhouette against the light background. This has probably some function in the mechanism of flight, when the parachute is borne away by the wind. If the plants are cut before the seeds are ripe, the lustrous balls remain whole and beautiful indoors almost indefinitely.

The specimen on the right shows the glossy, white calyxes which remain after all the parachutes have been blown away.

Unlike most anemones, *Anemone tomentosa* flowers in late summer and autumn. It is a tall plant with numerous pale red flowers. Seed catalogs and books on gardening praise it warmly, and I believe that it is being cultivated more and more. Strangely enough, I have never seen mention of the fruit, and yet I think it may be said without exaggeration that the fruit is far superior to the flower in beauty. As

93 · ANEMONE TOMENTOSA

with all anemones, the petals soon fall, and shortly afterward the branched stems sprout small balls the size of cherries, with dark spots on a white background. The dark spots are seeds, which move apart day by day as the ball grows larger. Eventually the ball begins to disintegrate, and the tiny seeds sail away. This is seldom seen out-of-doors, for the delicate tufts of fine fibers are soon destroyed by wind and rain, but if the plants are cut early and taken indoors, their singular beauty can be enjoyed almost indefinitely if they are handled very gently and kept out of drafts.

The withered leaves of the autumn anemone are also interesting. The upper surface is a rich umber while the under surface is gray and furry. As they wither, the edges curl up, resulting in a pleasing design in gray and brown.

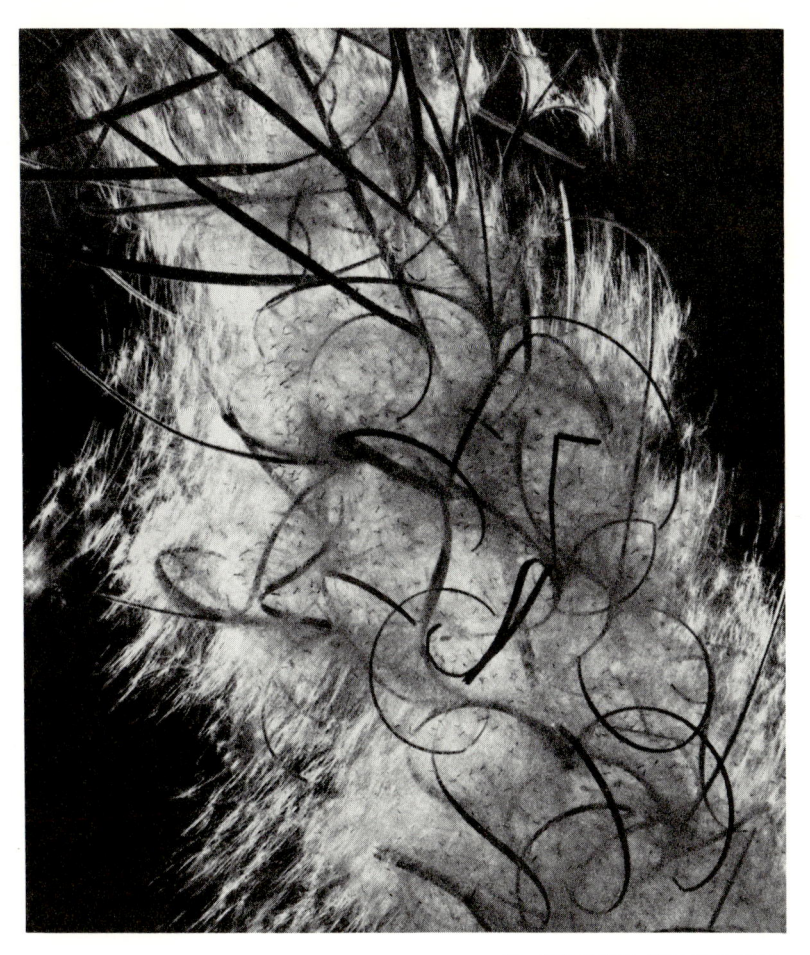

95 · EPILOBIUM ANGUSTIFOLIUM

*Epilobium angustifolium*, the Great willow-herb, has been neglected by poets and artists. It is difficult to understand why it is passed by in silence when many flowers far less beautiful, have been immortalized by poets. The plant is also called fireweed as it is an excellent ground cover and after forest fires, for example, it will quickly cover the burned area. Visiting London just after World War II, I will never forget the flower covered ruins in the city as Nature seemed to be doing her best to hide man's destructiveness under a mantle of one of her most beautiful flowers. The dense clusters of hanging buds redden during growth and fully developed, the rose-purple blooms in large masses are a glorious sight.

The flower stalk is really the seedpod, a feature which the botanist Gerard claims is responsible for an older name of the plant, Filius ante Patrem, "because the floure doth not appeare untill the cod be filled with his seed." The fruit is long and slender, and when ripe it opens into four elegant bows, between which the seeds display their downy wool. By pressing a pod gently between the fingers one can observe the beautiful opening ceremony. When a colony has ripened and revealed the downy seeds, it shines like silver, particularly against the light. The wind carries away the downy-winged seeds but the stems remain throughout the winter. They appear rather untidy, but on closer scrutiny attractive remains of the graceful bows may still be seen.

In the upper right-hand corner of the illustration is a pod beginning to open, while below it, a completely open one may be seen.

96 · THLASPI ARVENSE          97

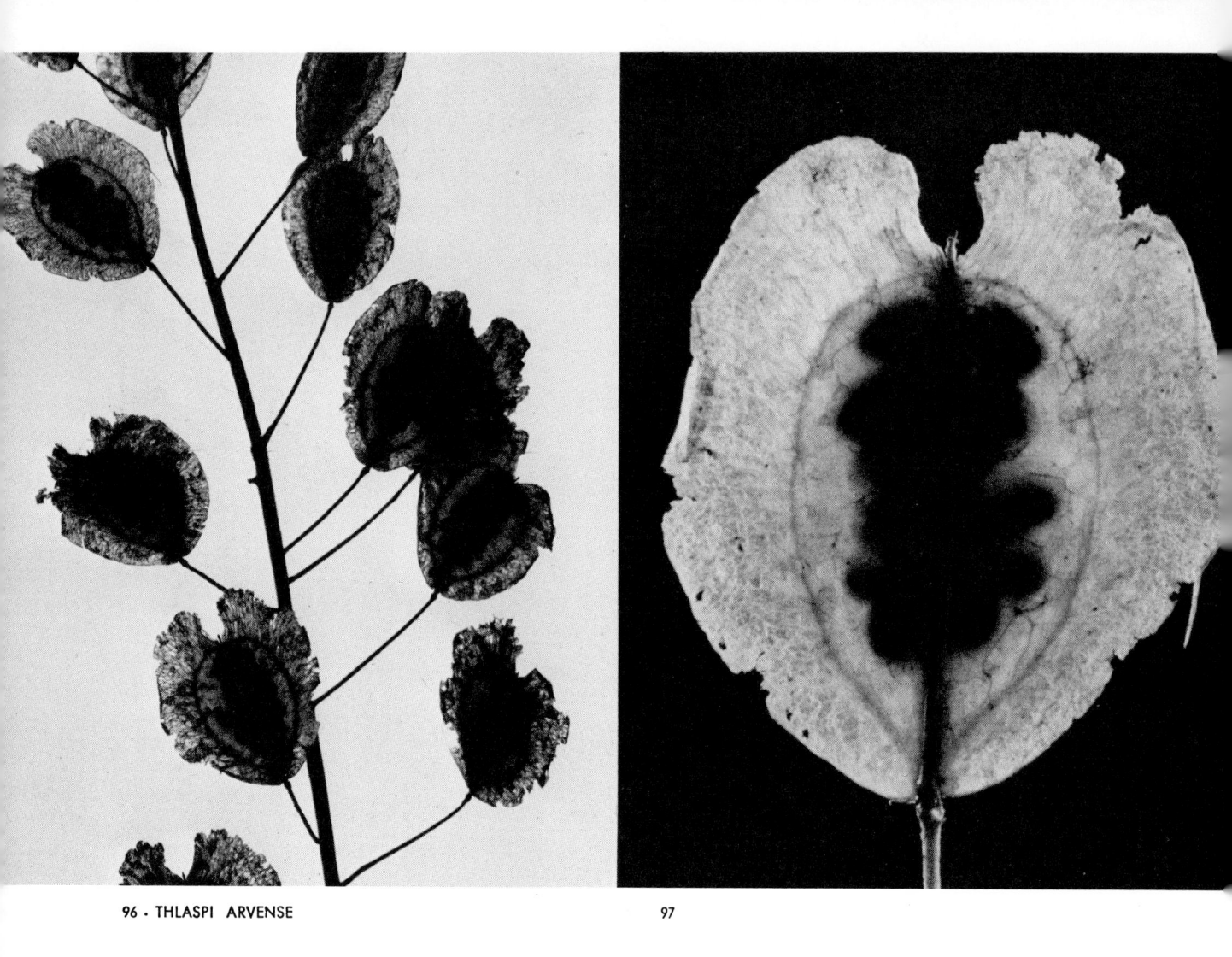

Pennycress *(Thlaspi arvensis)* has a reputation as a very hardy weed. An early 19th century botanist, simply called it "weed," and nothing more. Amateur gardeners groan at couch-grass, goosefoot, and pennycress and it is difficult to find anyone who has a good word to say for them. But as far as pennycress is concerned, I must say that it has its good points. Although not especially beautiful when in flower, it is attractive when it has grown tall and displays its many flat fruits. It is most striking when the seeds are ripe and the pods turn pale. Then, if seen against the light, the seeds may be discerned as black spots in the middle of the transparent pod, surrounded by a broad flattened wing, notched at the top.

98 · VIOLA TRICOLOR

The fruit of violets and pansies, little pale-green balls, look rather insignificant while they are growing, but when they ripen and open, the seeds pour out as from a cornucopia. The fruit bursts lengthwise into boat-shape sections, which are held together by the stem in the form of a three-pointed star, each section filled to overflowing with small, round pale-brown seeds. They pile up over the edges, and it is difficult to understand how the closed seedcase could have held them all. The explanation is that the oblong bowl containing the seeds exerts pressure on the walls from side to side. This is done so energetically that the smooth seeds are propelled a great distance from the parent plant. When the seedcase is empty, the collapsed walls are pressed tightly against each other. The picture above of *Viola tricolor* (wild pansy), does not do justice to the beautiful, seed-filled bowls; it is too dark for in reality the colors are pale and delicate.

101

99 . GERANIUM SILVATICUM

*Geranium silvaticum*, wood crane's-bill, got its name from the shape of its fruit. The scientific name, *Geranium*, is derived from the Greek *geranos*, a crane. A closer examination of the seedcase will reveal five ridges, each with a swelling at its base forming a small bowl containing one seed. When the seed is ripe and the time has come for the baby in its cradle to leave the mother plant and start an independent life in the world, the seedcase splits along the ridges. The five carpels roll up rapidly, and the seed in each bowl is ejected like a stone from a sling. The carpels continue to roll up and finally we have a five-armed candelabra, as shown in the illustration. At the end of July the plants are ripe and full of tiny candlesticks but as they are extremely delicate and break easily, many different varieties will be found on one plant, with from one to five arms.

100

103

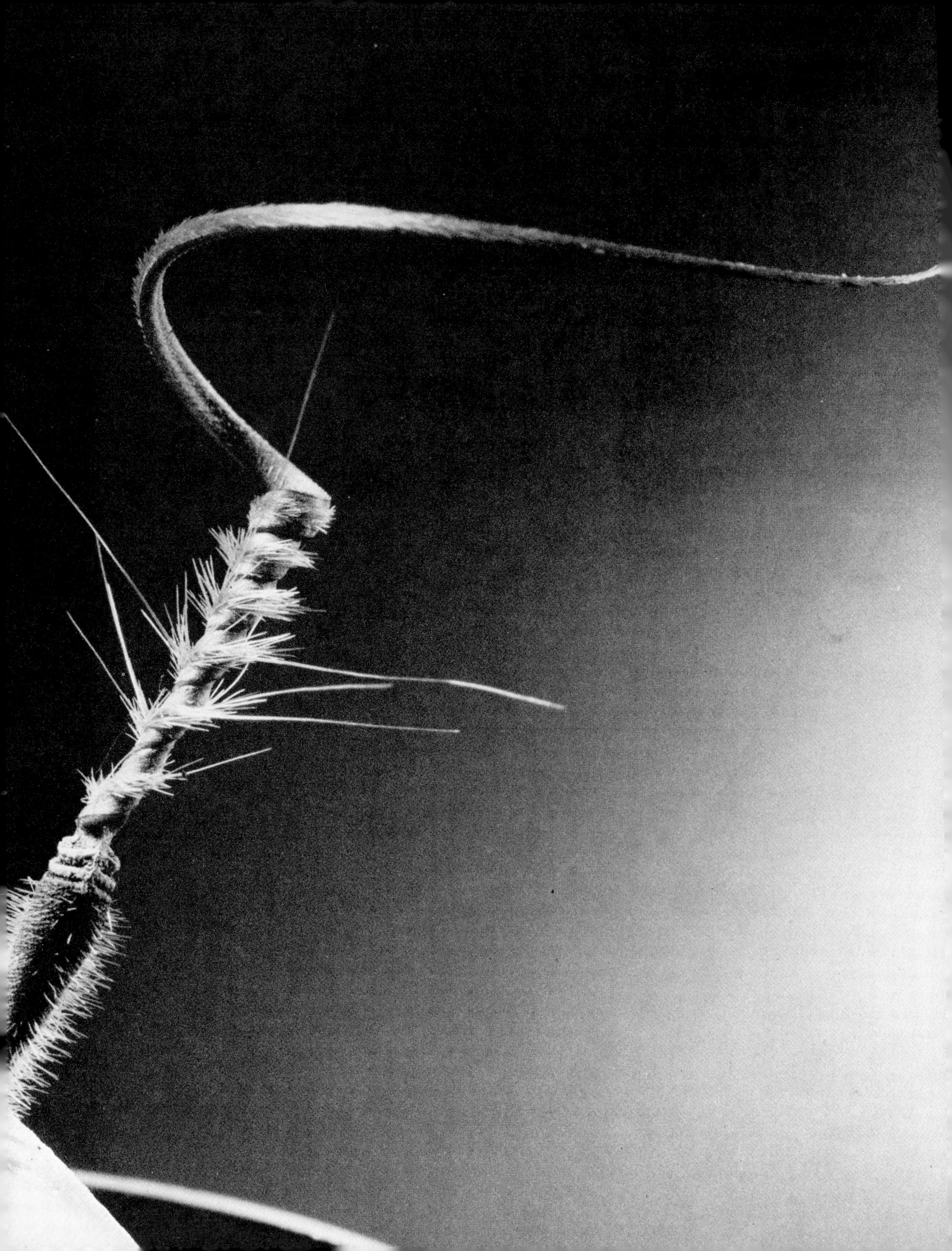

102

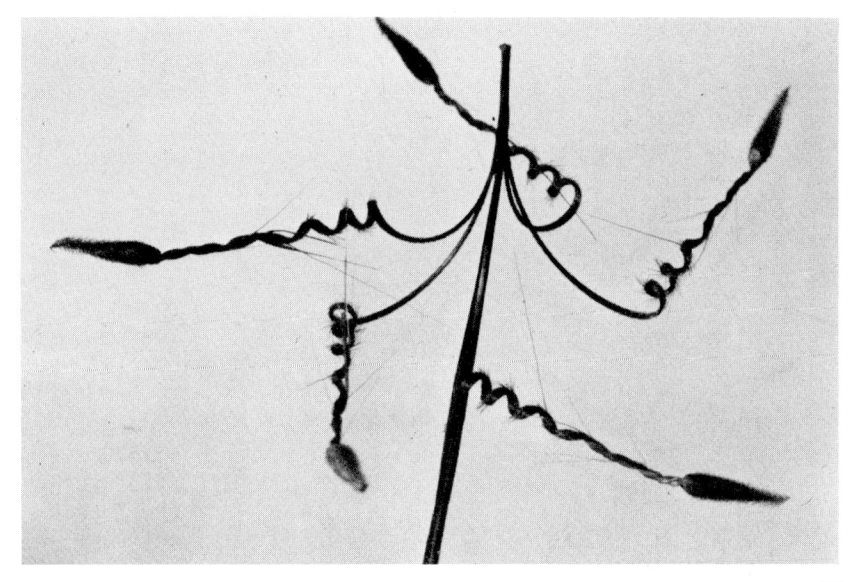

103 · ERODIUM CICUTARIUM

The English name of *Erodium cicutarium* is hemlock stork's-bill, but the Greek name more correctly translated is heron's-bill. Both names refer to the long seedcase. The plant belongs to the Geraniaceae, and is very similar to the crane's-bill but its seeds are dispersed in quite a different manner. Crane's-bill seeds are thrown abroad by the carpel rolling itself up vigorously in an elegant arch, while in the stork's-bill, the tails of the carpels with the seeds attached to the ends, twist into perfect spirals when they are released from the center pillar. This spiral is a marvel of engineering. It is very sensitive to moisture, untwisting when damp. When the seed reaches the ground, this untwisting movement helps to bury it deeper. Short hairs on the seed and long rigid hairs along the spiral, all backward-pointing, also play a part in this singular machinery, by braking and holding the seed, and ensuring that the spiral movement again drives it deeper into the ground. Two species of *Erodium* are shown here; *E. cicutarium*, the hemlock stork's-bill, is represented by the three graceful spires, each supported on five feet, which are the seeds, waiting for the moisture signal, to curl up in spirals; the other two are the fruit of *E. gruinalis*.

Occasionally the seed spirals remain for a time on the tip of the central column, and the plant resembles a curious candlestick. If one of these is to be kept in this form, a drop of glue should be applied to the top of the column. The seeds will then roll up as usual, but the glue will prevent their falling off.

101 · ERODIUM GRUINALIS

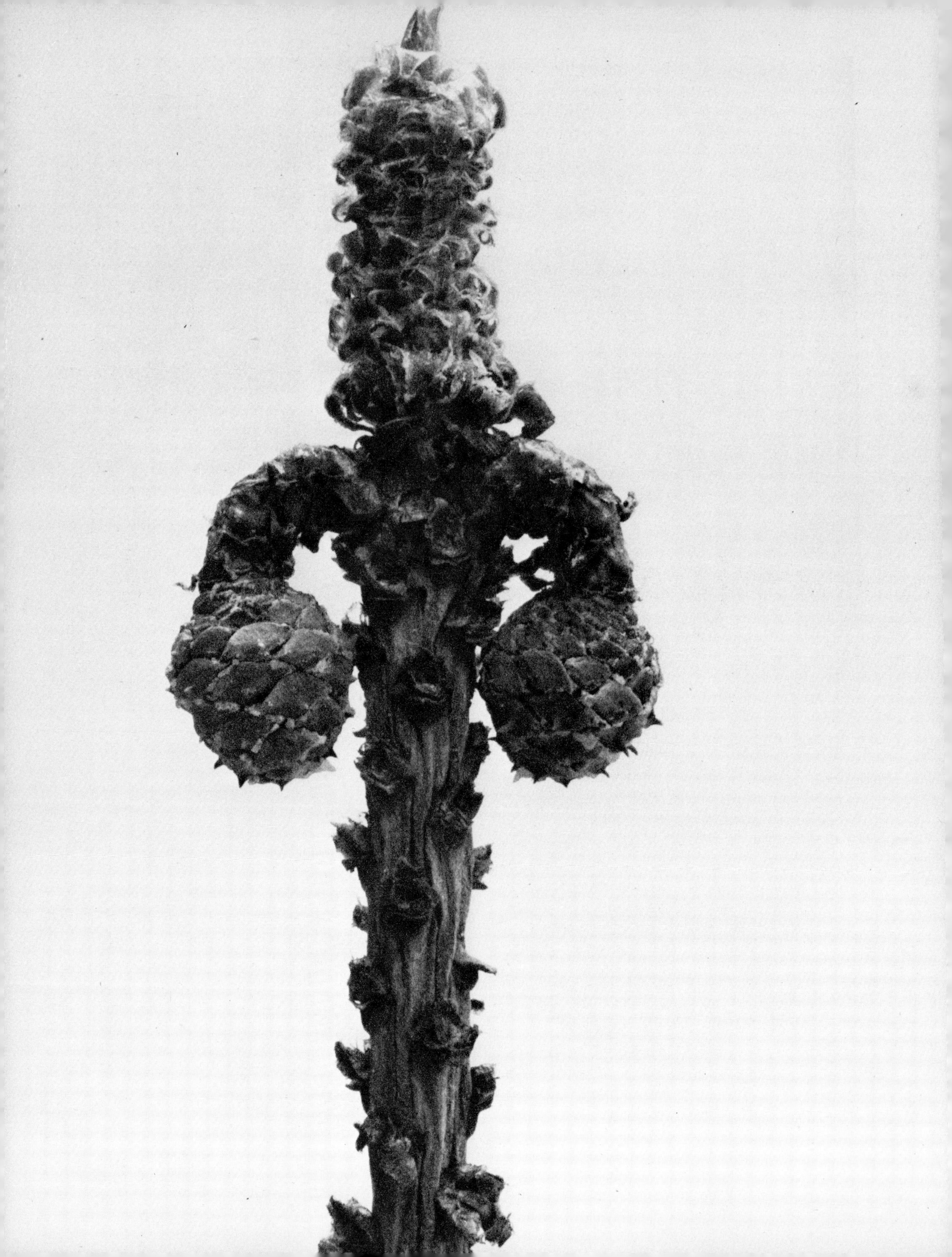

104 · PINUS SILVESTRIS

I know that there are many people who have never seen a pine *(Pinus silvestris)* flower, although everyone can recognize a pine cone. It belongs to the pine; it is there and that's that. But how it got there is something they know nothing about, and yet one of the most exciting things in nature is the development of a pine flower into the fruit. If you see a healthy young pine tree, a few yards high, at the right time, you will find a red lustre all over the tree. The young female flowers have opened. They grow like small clubs at the top of straight shoots with short needles, glossy and greenish-white (Illustration 106). The flowers do not remain upright. Soon after fertilization they bend their stems downward in an arch, the red color disappears, and the flowers harden to form the first cone phase. Often there are two flowers together, and when they bend downward, together with the top shoot they create a symmetrical formation, something like a fleur-de-lis (Illustration 105). Strong shoots may produce four or five cones.

107

108

Simultaneously with the red female flowers, cream-colored, occasionally reddish-yellow, pyramidal spikes of the male flower appear (Illustration 104). A cluster of needles begins to show or has already grown a few inches, at the top. When the male spikes are in full bloom, a sharp blow on the branch will send up a cloud of pollen which settles as fine powder on the surroundings, or is carried away by the wind. The small flowers in the spikes are arranged more or less spirally. Toward the end of the first summer, the cones are almost as round as eggs, with a distinct mosaic pattern. The following summer, they lose their childlike chubbiness and grow into pointed cones, of a dark green color, which becomes grayish-green and then gray, in the autumn. The seeds lie, well protected, between the scales of the cone during the following winter, and not until the third year does the mosaic begin to crack, when the sun warms it. The scales open wider and wider, leaving the way clear for the winged seeds. When the wind blows through the crown of the pine, the seeds fall out and are carried away by the wind to seek a place to grow.

The cones collected over the years become delicately colored. The pale-brown upper side of the scales, against which the seed and its papery wing have lain, has a narrow black border, and is equally dark on the lower side. Very old cones are of a distinctive silky-gray shade. The bottom surface of these old, wide-open cones becomes almost flat, and its mosaic pattern is like a precious jewel.

109

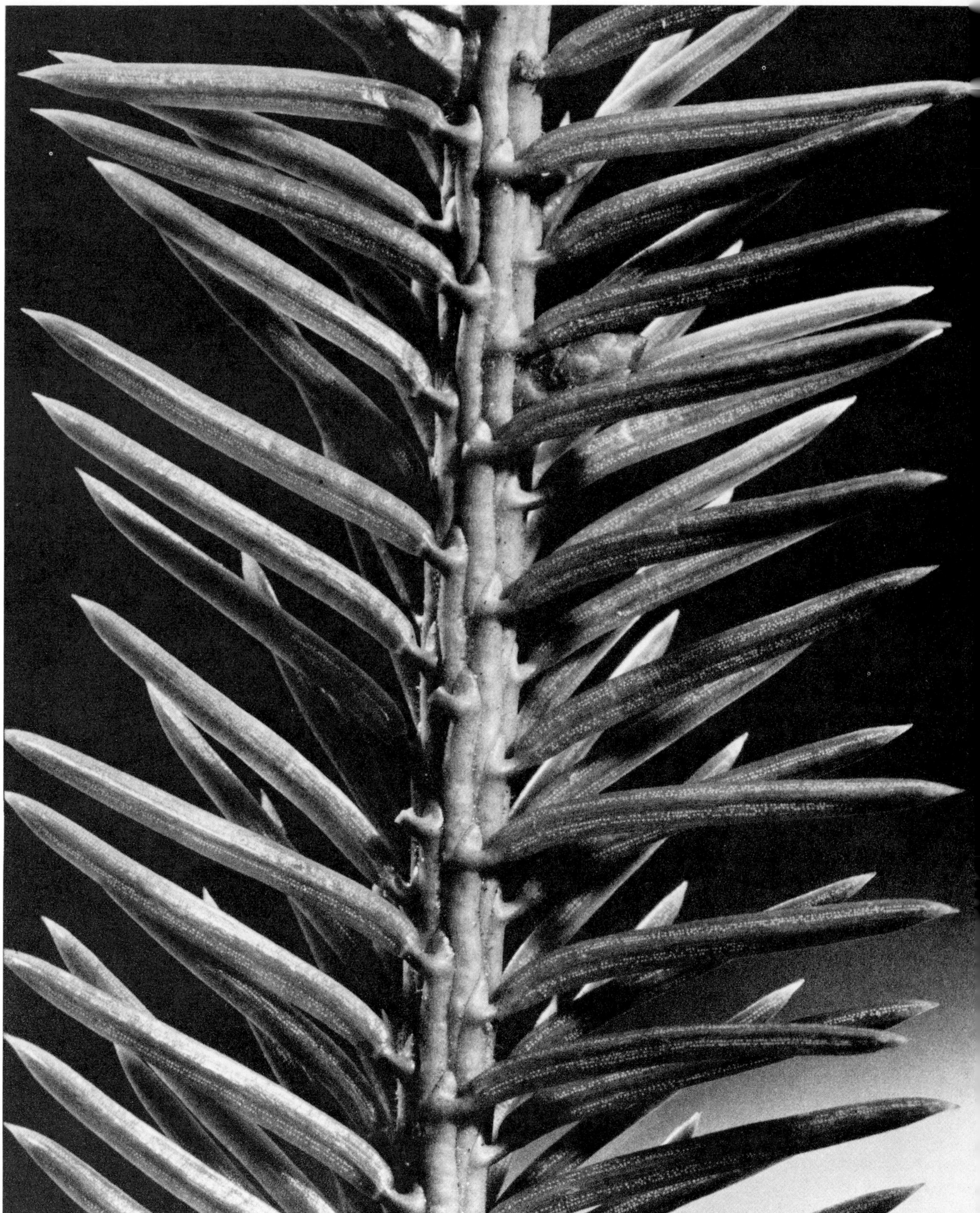

112

110 · PICEA ABIES

The spruce (Picea abies) has as many interesting details as the pine. The young, red female cones exceed those of the pine in both size and brilliance of color. The spike of the male flower is more modest in shape, but the amount of pollen it produces is incredible. Of particular interest is the way spruce needles are attached to the stem. On young shoots they are spread out on short, slightly bent stems emanating from ridges of bark. Together, these ridges form a longitudinal pattern. On the picture showing this pattern, the pale decor of several rows of fine dotted lines can also be seen. The needles of the top shoot are longer and stronger than those of the twigs; they stand upright, pressed gently to the stem, not unlike a number of homemade candles (Illustration 112).

The needle-less trees and branches of old windfall trees are patterned in a way which varies very much in branches of different size, but in principle, the patterns are the same. The points of attachment are paler than their surroundings, at times almost luminous, rhomboid in shape, and raised above the surrounding parts. Each point of attachment is like a little bracket for its needle (Illustration 110).

113 · ARAUCARIA COLUMNARIS

Fossil finds show that conifers belonging to the Araucariaceae family once grew in all parts of the world. The surviving species are now only found south of the equator. The family was named after the Indians of Arauco in southern Chile, for whom the edible fruit of one species was a staple food. Beautiful species of *Araucaria* are often cultivated in our homes and hothouses, and large trees flourish out-of-doors in southern Europe. The picture above shows branches of *Araucaria columnaris*. It is native to New Caledonia in Polynesia. I found it in the botanical garden at Coimbra. It had been windy the day before, and many long cones lay beneath the tall trees. Their scales are like curved bird's bills with broad bases. They are in regular spirals, like the scales of our spruce cones.

When I first saw the curious formations on a juniper bush, illustrated, enlarged, on the opposite page, I was rather astonished. They looked like fruit, but that was impossible, for I knew what juniper cones look like. I consulted my books, and at last found an explanation in Hartman's flora. These elegantly designed green goblets, which terminate in three pointed lobes, are a type of gall, corresponding to gall-apples on other bushes and trees. If one of them is cut through lengthwise, a tiny yellowish-red grub, the larva of a gall-fly will be found. Linnaeus knew them well, and named it *Hormomyia junipera*. In his account of a journey in Västergötland he mentions these galls as follows: "The galls were used as medicine by peasants here. They were collected and fed to swine, but God alone knows for what use and from whom they learned this medicine, for the common people seldom understand subtleties or the smallest insects. These galls are what the people of Småland call whooping-berries, and use to help their children when they have whooping-cough; they grow at the tips of juniper twigs and consist of three thick scales or needles, almost parallel, but opening outward at a point near the top. Inside are three small scales, joined together, containing a little red worm, which becomes a Tipula."

114 · JUNIPERUS COMMUNIS

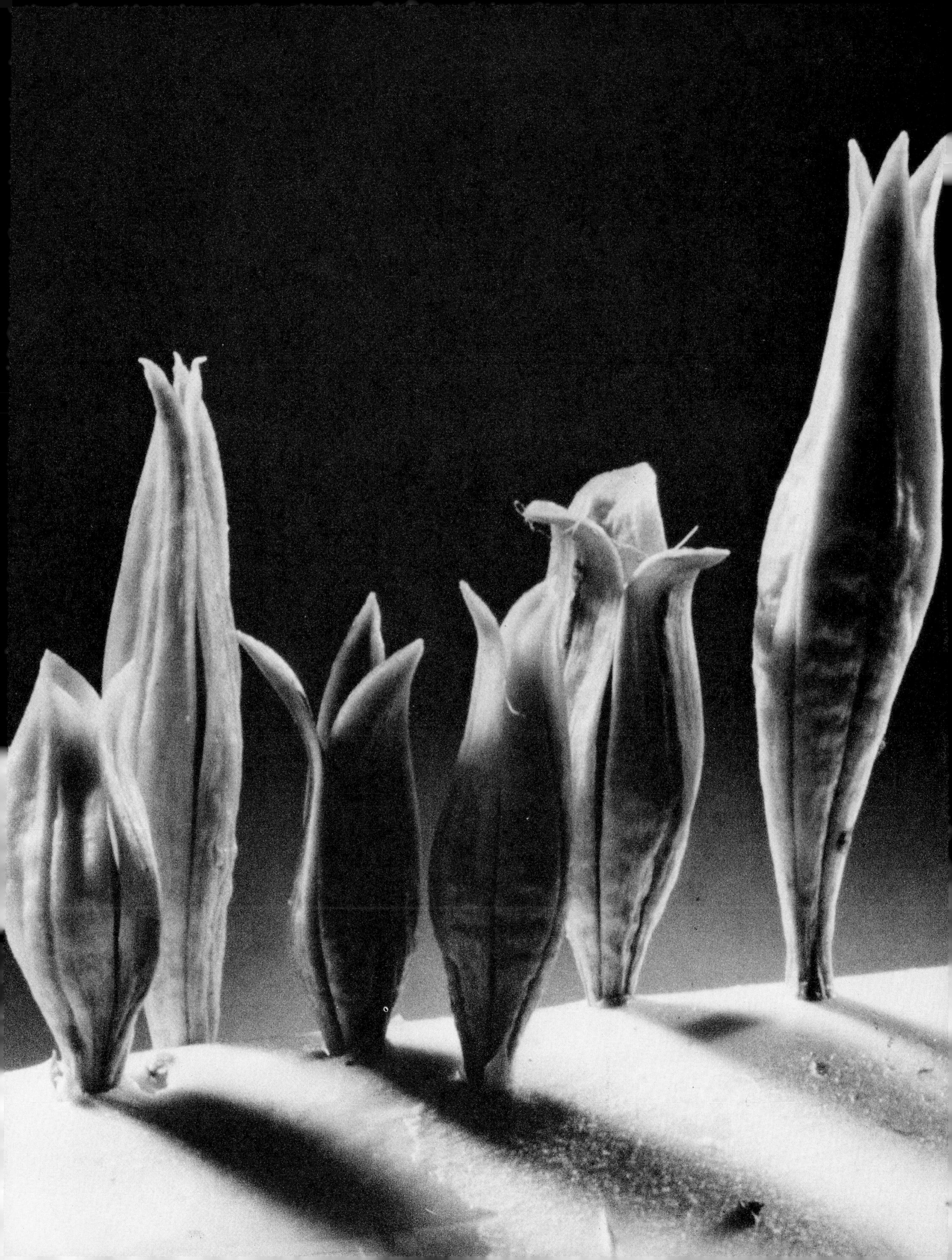

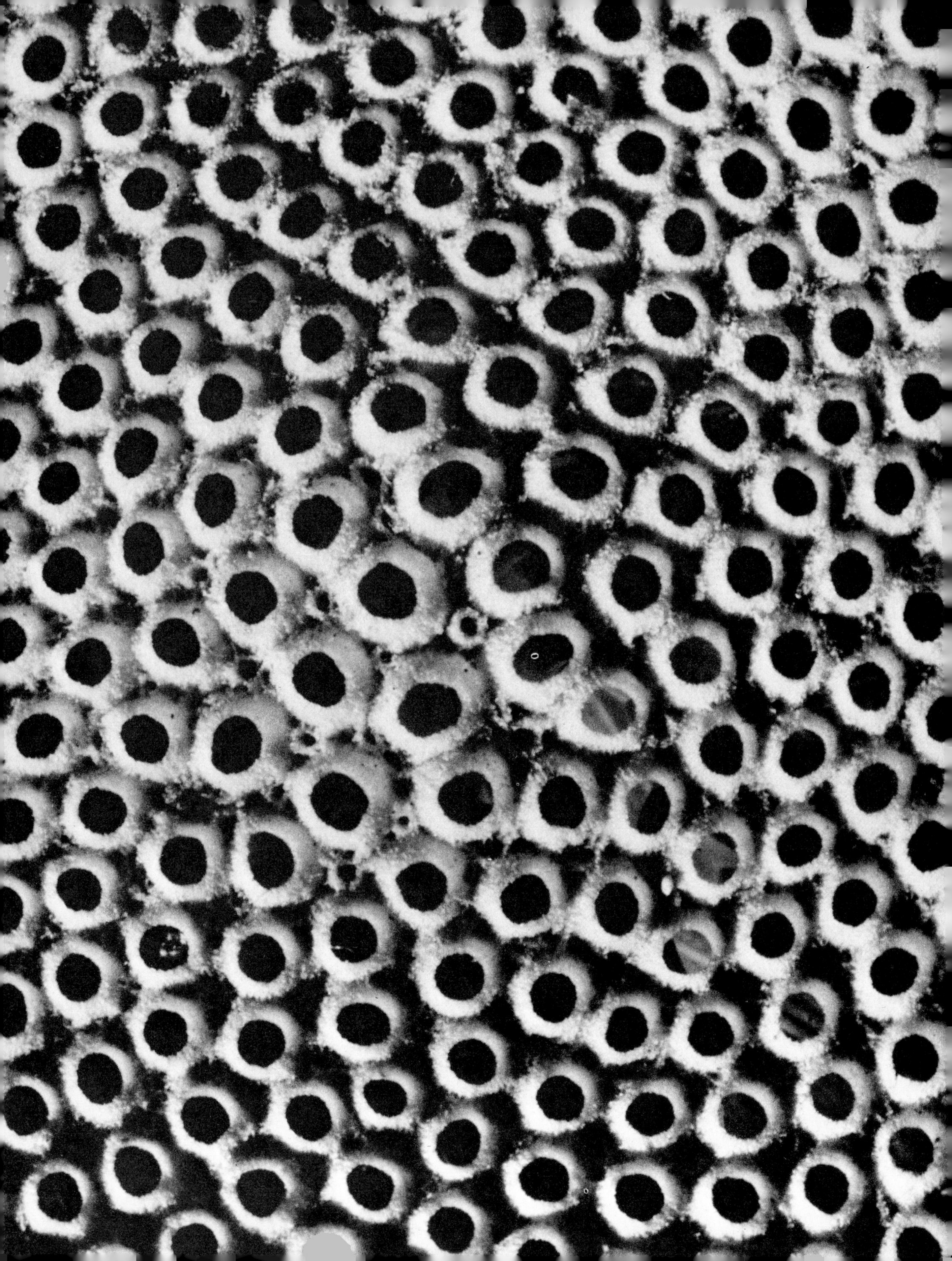

116 · CIRSIUM LANCEOLATUM

When late in summer, one wanders over fields and meadows where cows have grazed, one often finds the spear thistle's (Cirsium lanceolatum) gray-green, thorny bush. When thistle seeds ripen, it is as if a cloud of cottonwool had settled on the plant. Occasionally the tips of unopened plumes fasten together. It may have been a spider's web that fastened them together, or florets which matted together and had been unable to fall apart. When the wind blows, these brushes loosen and are carried off in a clump.

The picture above shows such a situation. The seeds have been torn from most of the parachutes, only a few in the lower right-hand corner remain.

The illustration opposite is of another species of thistle, *Carlina acanthifolia*, which is cultivated for its beautiful flowers and leaves. When all the seeds have left the plant in autumn, the entire receptacle falls off like a disc with the regular pattern clearly shown in the enlarged photograph.

115 · CARLINA ACANTHIFOLIA

117 · HELIANTHUS ANNUUS

Only a small part of the gigantic flowerhead of the sunflower *(Helianthus annuus)* can be shown in a photograph if details are to be distinct. Only about one-tenth of the flower can be seen here. The ray florets have fallen and the green sepals have bent inward over the edge to form a background of light triangles. The large sunflower seeds are lined up in long, undulating rows. Many are covered still with the remaining disc florets. The light spots on the seeds are the points where the petals were attached.

118 · DORONICUM CAUCASICUM

Leopard's bane *(Doronicum)* is often the first of the many perennials to bloom in our herbaceous borders. The flowers, like yellow daisies, appear in gorgeous clumps. The picture above was taken in late summer. The ray florets and disc florets have withered and the seeds are ripe. Some are about to sail away on their white tufts of hair. When they have all gone, only the bald receptacle remains, a cupola patterned with tiny depressions where the seeds had been attached. The seeds are a rather unusual shape; with a flat base and a goblet-like top, somewhat reminiscent of an old-fashioned stethoscope, and can clearly be seen in the detached seeds.

117

119 · ATRACTYLIS CANCELLATA

The thistle is the national flower of Scotland, probably because of its sharp prickles. The most distinctive order of chivalry in Scotland, the Most Ancient and Most Noble Order of the Thistle, has the motto, Nemo me impune lacessit (No one provokes me with impunity). But the Scots were not blind to the beauty of the plant, which is often taken as a pattern in works of art. I do not know which thistle was chosen as the symbol of Scotland, but most thistles are well armed, and all are very beautiful. The illustrations here show a species which takes the prize for beauty. It is a small plant, *Atractylis cancellata*, from a dry, sandy hill in southern Portugal. It is in fruit, and the two seeds in the left-hand picture are just about to start off into the world on their fine tufts of hair. A cage of pinnate leaves around the flowerhead is completely transformed into thorns, an airy, graceful creation of enchanting beauty. The word *Atractylis* is Greek, and is the name of a thistle, used as a distaff in spinning; the species name, *cancellata*, means, very appropriately, "surrounded by a grating." The right-hand illustration shows this grating after the flower has been removed.

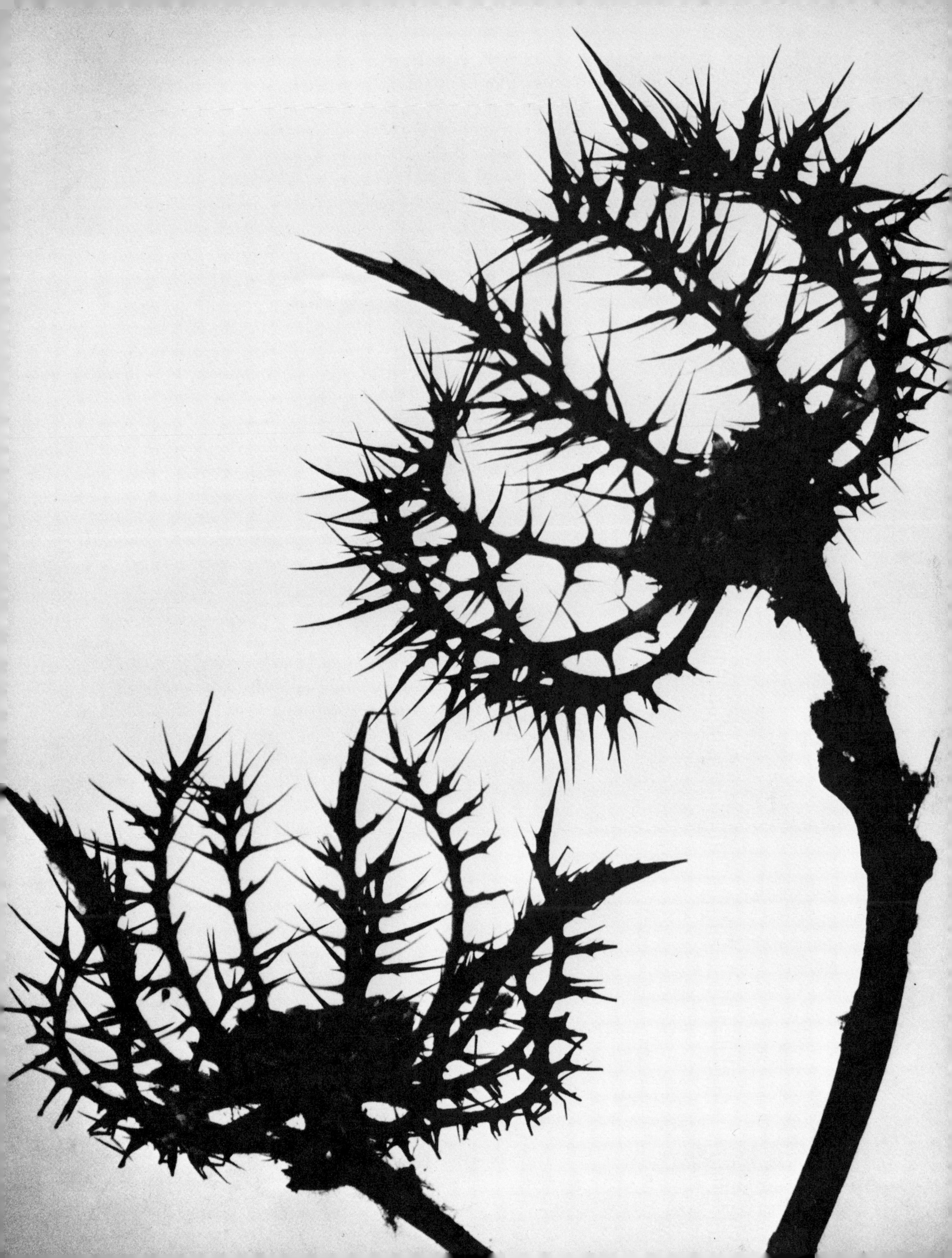

121 · RUDBECKIA PURPUREA

The flowers of daisies and allied plants often have a stricter, more stylized form before they are full-blown. Later the contours become rounder and softer. Here is such a flower in the late bud stage. It is a tall perennial, often grown in borders, *Rudbeckia purpurea.* The rosy ray florets stick up like a palisade around the tightly packed pointed bullets in the middle; a compact and firm composition. In full bloom, the ray florets bend backwards in graceful arches.

The shape of the receptacle of plants belonging to the Compositae varies greatly from species to species. It may be flat or round, concave, convex, conical, cupola-shape or almost spherical. The forms of the florets and their fruit are determined by the shape of the receptacle. One picture shows the fruit phase of the scentless mayweed *(Matricaria inodora)*, after the white ray florets and the yellow disc florets have fallen. Each little square in the mosaic pattern was once the base of a floret and the top of a seed. When all the seeds have dispersed, there remains only the cupola-shaped foundation, the receptacle. The *Rudbeckia lanceolata* shown in the photograph opposite has reached this stage.

122 · MATRICARIA INODORA

123 · RUDBECKIA LANCEOLATA

124 · ARCTOSTAPHYLOS UVA-URSI

The plant with the impressive name of *Arctostaphylos uva-ursi* is called bearberry in English. Both the family name and the species name mean bear-grape. The berries are dry, almost mealy in consistency. The plant may be found growing in moors, and on stony mountain heaths, in large creeping carpets, extending long trailers. The small, glossy, bright green leaves keep alive all year round. The bearberry is closely related to the bilberry and the cowberry, and the petals of the flowers are joined together, gamopetalous, forming a bell, which falls when the seeds begin to ripen. The picture above shows two such corollas with their characteristic shape, something like the chimney glass of a kerosene lamp. When the flowers have fallen, the twigs have the appearance shown in the illustration opposite. The ovary reminds one of a Christmas-tree candle.

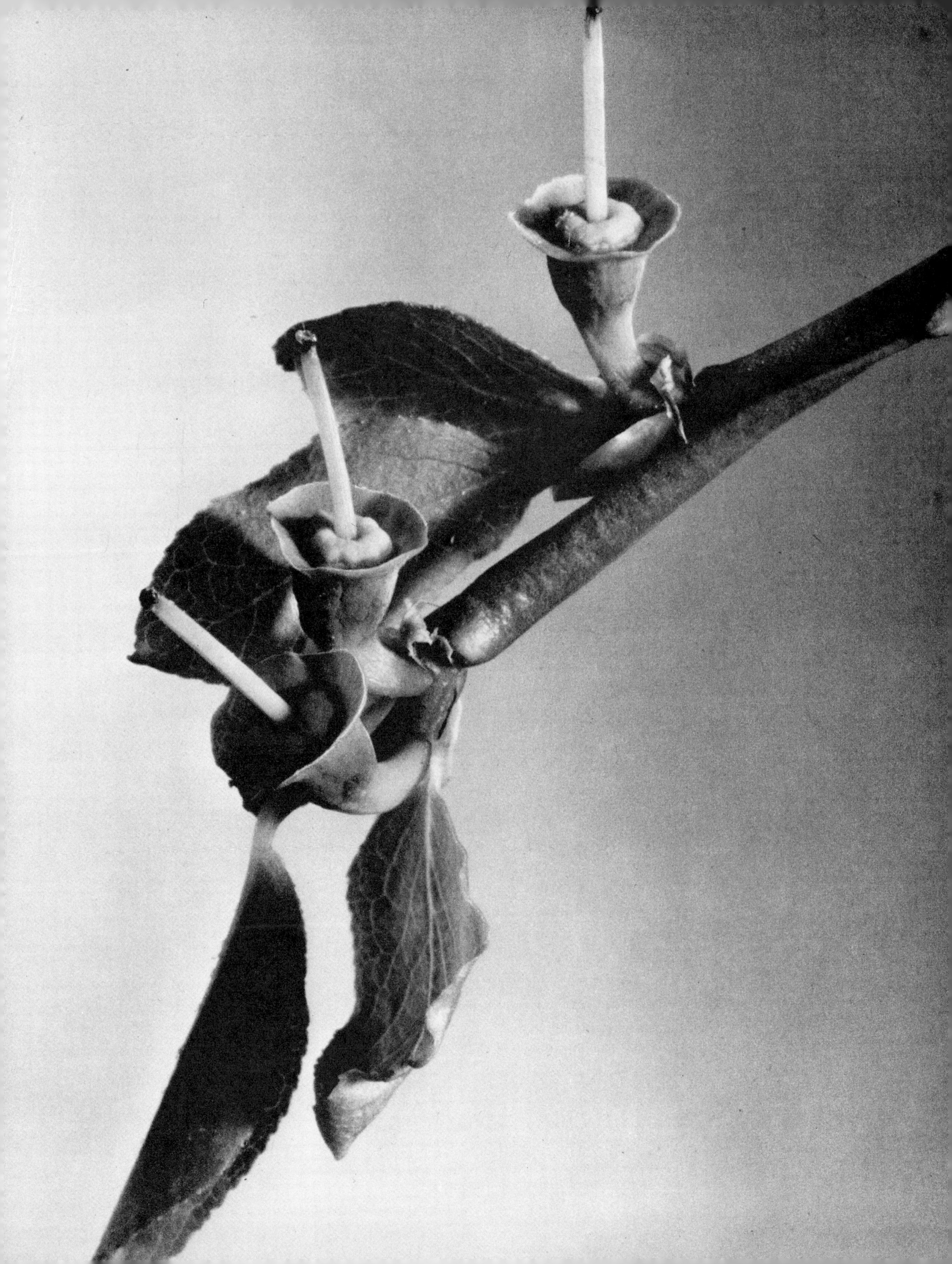

126 · PYROLA SECUNDA

*Pyrola secunda.* The sixteenth century French botanist, Charles de Leclus, gave the beautiful name, *Pyrola*, to the plant shown above. It means little pear-tree, an appropriate name, for this evergreen plant which the English call wintergreen, looks like a little tree with leaves like the pear-tree and fruit somewhat like flattened Bergamot pears. The five follicles are bound by deep furrows, and the pistil remains suspended for a long time from the middle of the fruit, much like an inverted candle.

When the purplish-blue flowers of Self-heal (*Prunella vulgaris*) have withered and fallen, the formerly irregular flowerhead becomes a formal and concentrated composition. The hairy calyx is two-lipped; the outer or lower, with two pointed lobes, is seen against the inner lip. The flowers are in whorls around the stem, one above the other, forming a dense, terminal head. Each whorl is enclosed in or supported by a very beautiful bowl-shape formation of leaves which can best be seen at the base of the flowerhead. They remind one of the dish-like leaves of the honeysuckle.

127 · PRUNELLA VULGARIS

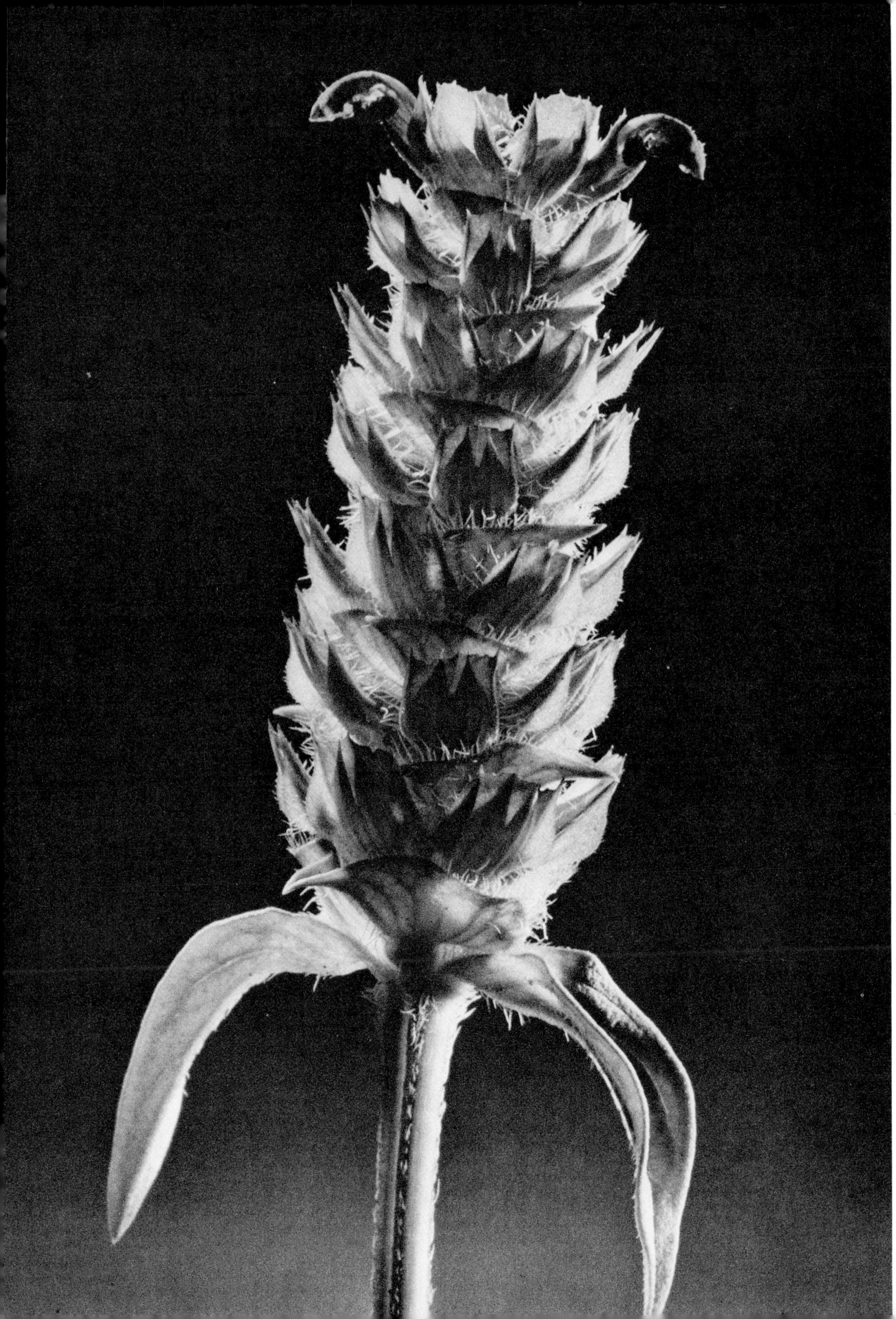

128 · SALVIA VIRIDIS

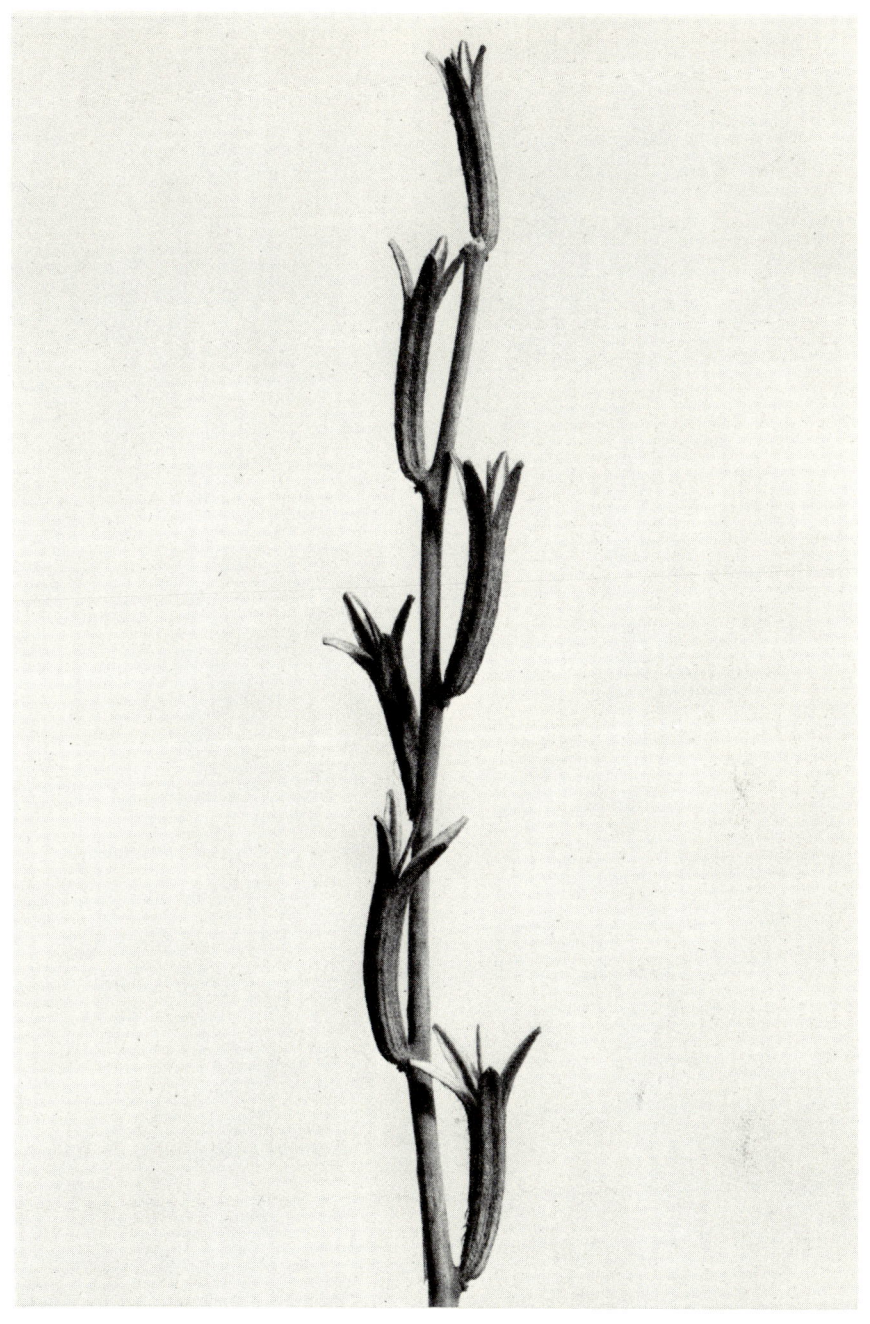

129 · OENOTHERA GRAVEOLENS

At first sight, the pictures appear to be composed by a designer whose object was to supply modern lighting units for a large concert hall, or to design some ornate ironwork. The flowers of the Sage plant opposite, *Salvia viridis*, have withered and the seedcases are turned downward. The other photograph shows the fruit of *Oenothera graveolens*. In a spiral, slightly curved, they are thinly distributed along the long stem. Only a half-dozen are shown in the illustration; they have just begun to open to release their tiny seeds.

127

130 · TREMATOLOBELIA MACROSTACHYS

**131 · LOBELIA GLORIA-MONTIS**

*Lobelia erinus*, a common edging in our gardens is an annual plant, quite small, with numerous small blue flowers. There is a wild lobelia in England, *Lobelia dortmanna*, which grows on the gravelly bottoms of shallow lakes in the west of England. Many other lobelias, both trees and shrubs, are found in other countries. The fruit shown opposite, of *Trematolobelia Macrostachys*, belong to a plant closely related to the lobelia. It was given to me by Professor Skottsberg, who brought it home from the Hawaiian Islands. It is a tree lobelia with large flowers and leaves like rosettes. The fruit is rather similar to those we find on our campanulas, with window-like openings through which the seeds are thrown.

*Lobelia gloria-montis* grows on the slopes of African volcanoes. Seen in the small illustration are two of this species which is also found as trees. To judge by the name, *gloria-montis*, which means glory of the mountain, it must be a splendid plant.

132 · ASTRAGALUS FALCATUS    133

*Astragalus* belongs to the Leguminosae, the pea family. Some species are cultivated for their beauty, others, such as the species illustrated here, *Astragalus falcatus*, grow wild in eastern Europe. I was attracted by the decorative clusters of fruit which are strikingly like the spruce trees in the drawings of Carl Fredrik Hill.

134 · LUNARIA ANNUA

*Lunaria annua* has various names in English. The botanist, Gerard says, "We call this herbe in English, Penny floure or Mony floure, Moon-wort, Satin and White satin and among our women it is called Honestie." The name Honesty is the one we use, for all its character is clearly visible. It is the satiny film between the two halves of the flat fruit that has inspired many of its names. Before the outer walls of the seedcase have fallen off, the seeds, seen through the thin, transparent tissue as dark spots, are kidney-shape and obtain nourishment through a fine channel running from the edge of the pod, like an umbilical cord, to the seed.

The remains of these cords are visible in the picture as fine black lines. A few well-developed specimens of Honesty will keep almost indefinitely, and brighten up the dark winter days.

131

135 · SOLANUM DULCAMARA

*Solanum dulcamara*, called bittersweet or woody nightshade, is a near relative of the potato, imported from South America. It has a much smaller, though generally similar structure, and flowers of the same purple color. It prefers dampness and grows alone but likes to climb trees and other plants. The specimen shown above was found on a stretch of land always flooded in spring. The vegetation in the vicinity was luxurious, and bittersweet trailed over alder bushes, marsh cinquefoil, purple and yellow loosestrife.

The structure of the flowers, fruit and stem, is unusually graceful. The ripe berries, which are a bright red, remain on the plant for a long while but when they have fallen off, they leave a structure resembling a candelabra.

The plant's rather contradictory name, bittersweet, which should be sweet-bitter if the Latin name were translated literally, is due to the fact that at first the bark tastes sweet when chewed, but after a time it becomes bitter. Bittersweet was formerly used as a drug to treat eczema and gout, to mention only two of the many diseases it was reputed to cure.

Dyers Woad (*Isatis tinctoria*) has become less known in this age of advanced chemistry. Its leaves were formerly used to make a blue stain for dyeing cloth, and it is believed to be the color with which the ancient Druids dyed their bodies blue. To quote a sixteenth century translation of his *Commentaries*, Caesar wrote, "Al the Britons doe dye themselves with woade, which setteth a blewish color upon them ." Later woad was superseded by indigo, made from an East Indian plant, *Indigofera tinctoria*. Woad is still found growing wild in northern Europe, and is also cultivated as it is a beautiful and stately plant, with a many-branched stem and large clusters of yellow flowers. I have included it for the sake of its fruit. The flowers are in clusters around the top of the plant, but as the seeds ripen and become heavier, the long clusters droop on thin stalks in graceful arches, the leaves fall at the same time. When fully ripe, the flat pods are dark brown, almost black. During the transitional period, the color is a curious mixture of green and black, but the black spreads very rapidly over the entire surface.

The Pharaohs of Egypt were fond of flowers and their mummies were embellished with garlands and chains of flowers and leaves. I do not know whether woad grew in Egypt then, but I imagine that Amenhotep IV would have enjoyed seeing his beautiful wife, Nefertete, in a necklace of woad fruit.

136 · ISATIS TINCTORIA

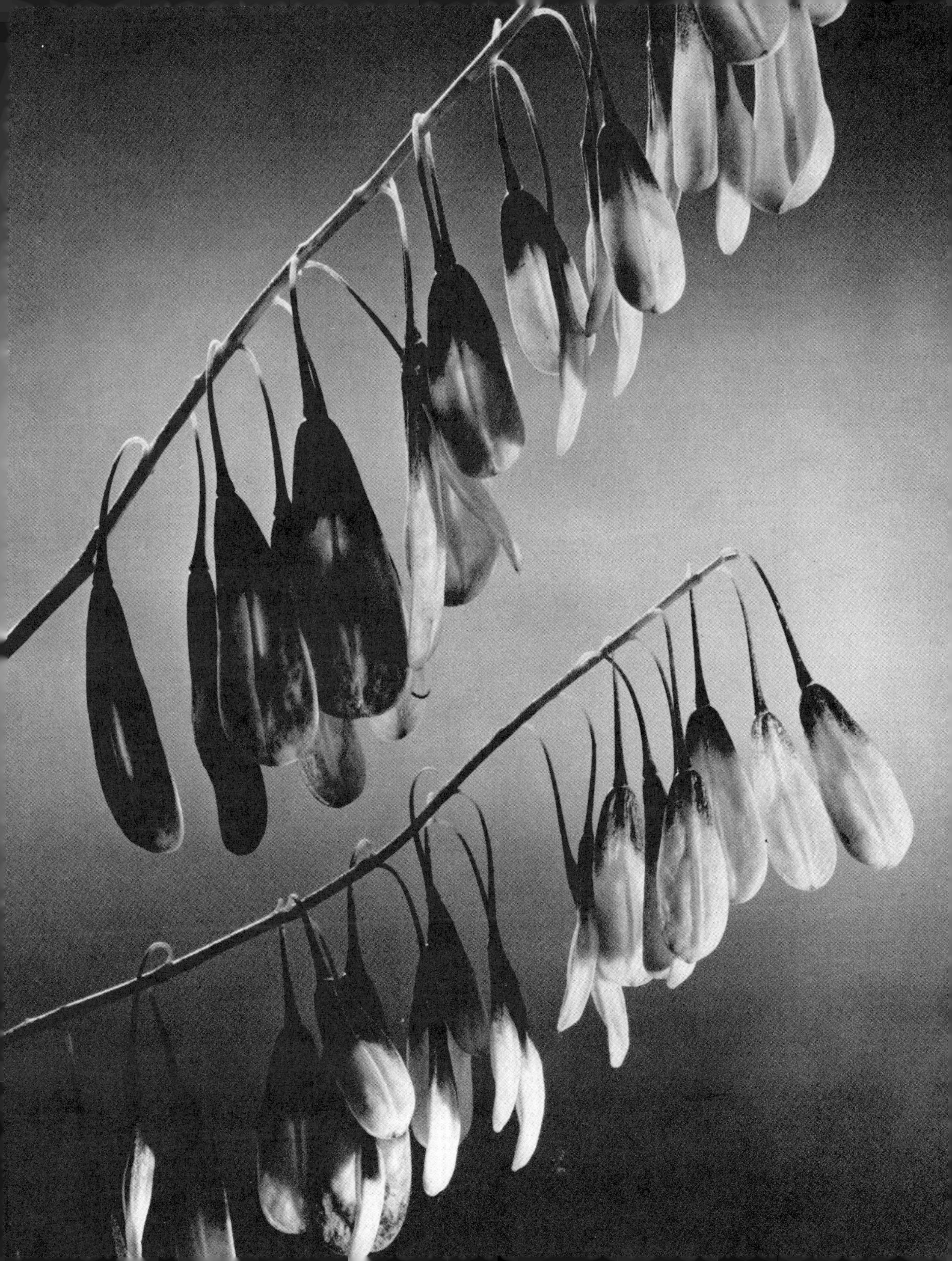

138

139

134

137 . CORYLUS AVELLANA

The beauty of the golden-yellow catkins attracts our interest to the hazel bush in spring, but gathering the delicious hazelnuts or filberts, draws us to the hazel thickets in autumn. The male flowers form these beautiful catkins with their clouds of pollen; the female flowers, far less conspicuous are, however, well worth looking for. They are tiny, bud-like formations, with a cluster of bright red stigmas.

The nut is surrounded by a leafy cup. In an old flora, I found this cup described as open and bell-like, with lacerated, toothed lobes. Perhaps the word lacerated had a different meaning in earlier times, but if it meant that the involucre is irregularly gashed, it is the wrong word to use, for the edge is uncommonly beautifully patterned like fringed lace, while the veins on the surface form a fine net. At times the involucre is rather simple, reaching to the approximate middle of some of the hazelnuts, while at others, it has numerous long teeth and envelops the entire nut, extending far beyond the point. Every bush seems to have its own peculiarity. The base of the hazelnut is radially striated, and the periferal circle line is marked inward by a row of fine teeth.

140 · PLATANUS ACERIFOLIA

While strolling along the magnificent Avenida da Liberdade in Lisbon, I tried to avoid a cloud of dust stirred up by a street sweeper, but at the same moment caught sight of the fruit pictured above. I managed to catch it just as it was about to roll through an iron grating. It is, I suppose, a plane tree fruit or more correctly, the receptacle remaining after the seeds have fallen off. The fine network, which has largely loosened from the surface, has left a mosaic pattern of shallow furrows. My eye had quickly reacted to the small, net-covered sphere, but when I looked around for others, I could not find any. I was standing under a plane tree, probably belonging to the species *Platanus acerifolia*, a cross between *P. orientalis* and *P. occidentalis*—a stately tree with a gray bark which flakes off in large, pale patches. In England it is known by the name, London plane; in parts of Portugal I saw gigantic trees of this species. They grow rapidly, and may become very old and thick. I have seen a report of a 100 year old plane tree in Ragusa, measuring more than thirty feet in diameter, and near Istanbul there was a plane tree with a diameter of about fifty feet. Gottfrid de Boullion, the leader of the first Crusade, is said to have rested under this tree more than 800 years ago.

Some years ago, while walking across an island with the setting sun behind me, approaching an open field I saw a lovely sight, the gleam of silver flowers on branched stems. At a distance it seemed as if the ground were covered with silver coins. I found that they were withered red cornflowers; the florets had fallen, and all that was left was the empty, gleaming receptacle illumined by the slanting rays of the sun. The blue cornflower illustrated here has a similar lustrous receptacle although somewhat smaller. The depressions in the center show where the florets and seeds were attached. The scaly exterior of the calyx is also worth studying.

136

141 · CENTAUREA CYANUS

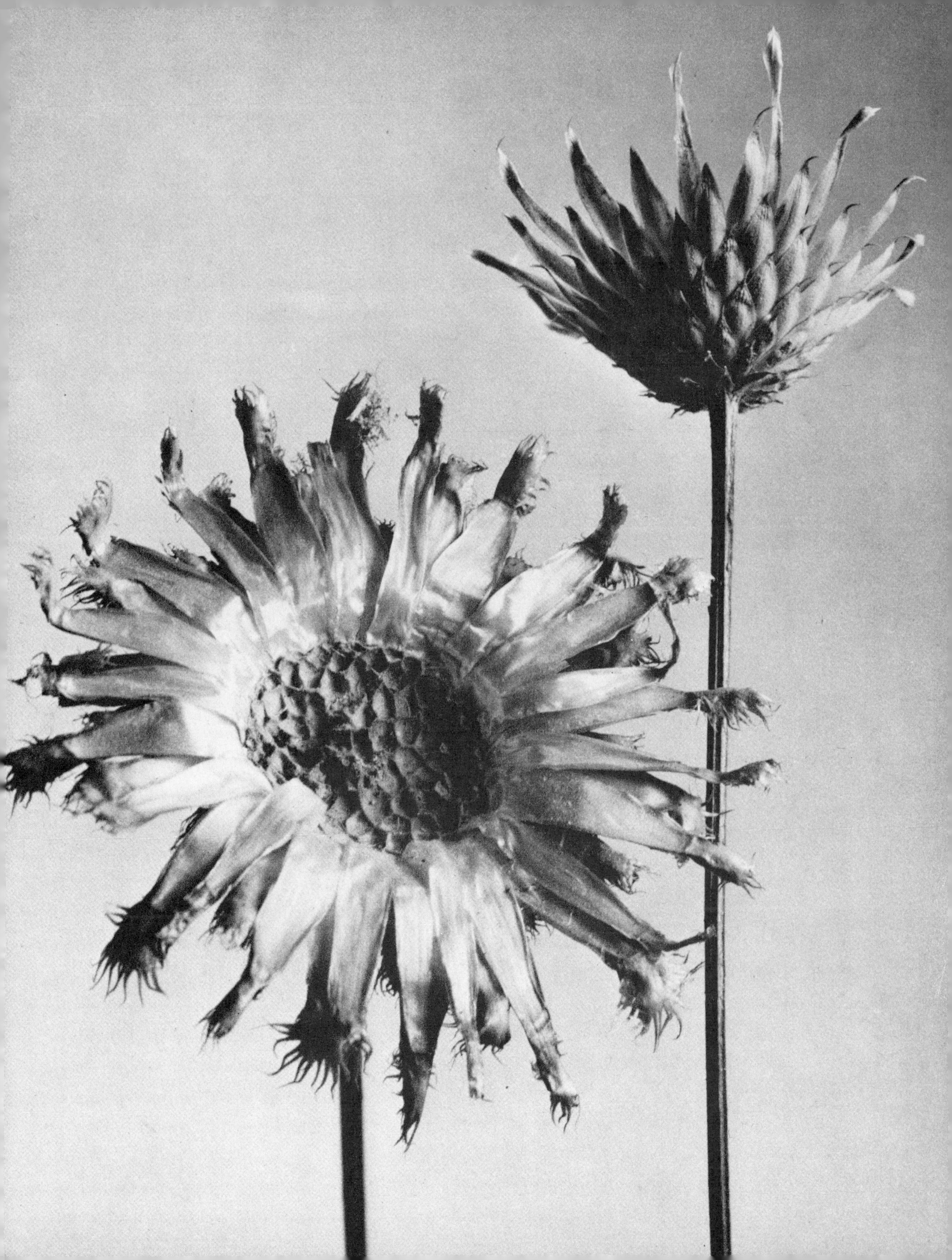

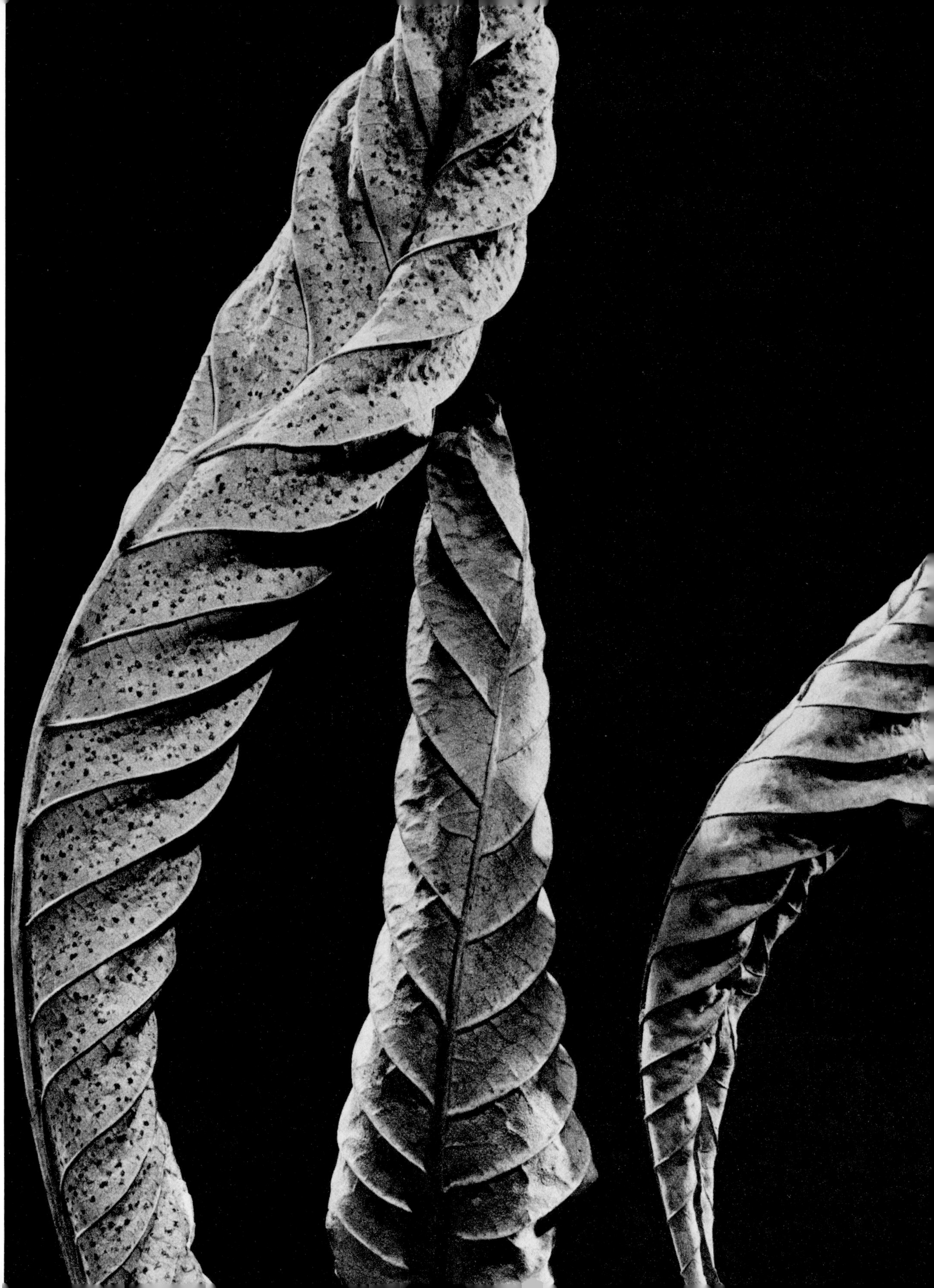

142 · AESCULUS HIPPOCASTANUM

A Swedish author, Björn von Rosen, wrote in one of his enjoyable books on nature that he carried Horse-chestnuts in his pockets because they are so pleasant to hold and touch. I have often done the same. Children everywhere invent games to play with them. English boys play a game called "conkers", drilling a hole through the nut, a string is threaded through and knotted. This makes a conker, which is swung at an opponent's and the winner will have smashed the other's conker. It is quite unnecessary to praise the Horse-chestnut tree as it is popular in any case for its magnificent leaves and the pale pyramids of flowers against the dense, dark green foliage just as spring turns into summer. Think of the sticky buds, so full of vitality, examine the detail of the flowers, the individual flowers in white and pink, and the clusters of stamens in S-shape lines. Finally, pick up and examine a few sun-dried leaves, wizened and patterned with the sweeping curves of the veins.

It is seldom one sees the tulip fruit. As soon as the petals fall, the stems are cut down to prevent the fruit from drawing nourishment from the tulip bulbs and spoiling the following year's blooms. Experience has shown that this is very wise, but one will not regret allowing a few tulips to run to seed after flowering. The petals fall first and the elegant stamens follow, then the rudiment of the fruit in the powerful pistil soon begins to swell. It stands like a mighty pillar, and perhaps the Dutch took the shape of the pistils as a pattern for their earthenware bottles, so pleasing to the eye. The pillar terminates in the remains of the stigma in small, wavy lines or ridges. Slowly the fruit develops into the final shape. Their color changes from green to pale yellow or brownish-yellow, the walls become hard and dry; then finally the fruit opens at the top, the seams burst and the walls fall apart. The surface is transversely striated between the seams joining the three or four walls. These striae correspond to the seeds, which lie piled one upon the other in their compartments like rolls of coins. Tulip growers are incessantly creating new varieties; a lovely pale-pink tulip just recently developed was named after the lovely Swiss opera singer, Lisa Della Casa. The fruits do not vary as much as the flowers, but still quite enough to make them interesting.

145

144 · TULIPA

*Bidens tripartita*, the three-cleft bur-marigold, which belongs to the Compositae, is rather dull in color; brown, yellowish-brown and blackish-brown. The flowerheads do not attract one's attention to any great extent, but when in fruit it has many interesting features. The flat, oblong seeds are edged with small, sharp barbs, and these barbs continue from the square-cut upper part of the seed, on two long bristles. A third barbed bristle is frequently found between them. Between the florets, and later between the seeds, are membranous, lancet-shape scales, which become pale and hard as the fruit ripens. When the flowers have withered and the seeds fallen, or have gone, fastened by their barbs on the fur of passing animals or clothing, these scales remain, spread out to make a glossy star. The first time I saw these stars on a damp slope between a barn and a sawmill, they sparkled in the evening sun and I was amazed to find that the shining display was only the remains of the humble three-cleft bur-marigold.

146 · BIDENS TRIPARTITA

147 · GEUM RIVALE

148 · GEUM URBANUM

Water avens *(Geum rivale)* is another of the plants whose seeds are dispersed with the aid of hooked styles by which the seeds are caught on the pelts of animals or on clothing. These hooks are at the ends of the long styles projecting from the oblong seeds. These styles are the hardened remains of the pistils, which are ingeniously contrived. During the early stages of the flower's development, the pistils, or more accurately, their stigmas, are divided into an inner and an outer part. The outer part resembles a soft feather and is connected by an S-shape joint to the inner part. When the seeds are ripe, the outer, soft part of the pistil falls off and the hook is released. I suppose this is nature's wise arrangement to prevent the seeds from being carried away before they ripen. The feather is discarded when it is no longer needed to protect the seeds. The illustration shows seeds both with, and without protection, for the hooks. Part of the water avens flower has been removed to show the details more distinctly. When all the seeds are gone, a soft, downy shaving brush remains, the axis on which the pistils and stamens had been supported.

Herb bennet *(Geum urbanum)* has fruit of the same type as those of the water avens, but the outer style is much shorter and only very slightly hairy at the base. Generally, in the water avens, the hairiness of the fruit is much more pronounced.

The botanical name of the burdock was formerly *Lappa tomentosa* (now *Arctium tomentosa)*, and it gave a good

149 · ARCTIUM TOMENTOSA    150

description of the plant. *Lappa* is derived from the Greek *"labein,"* to hook fast, and *"tomentosa,"* meaning downy. The so-called involucral bracts taper from a base to a bristle with a very powerful hook. The downy effect is due to fine, cobwebby threads spun tightly around the entire flowerhead and between the hooks. Each time one sees them, a number of questions occur. Man invented the fishhook, but what forces gave rise to the burdock's hooks which are so useful to the plant? And what is the purpose of the cottony down? Closely related species have no such down and yet they seem to flourish.

In the left-hand picture above, the young flower buds can be seen through a hole made in the calyx. They are very well protected during growth by the firm scales.

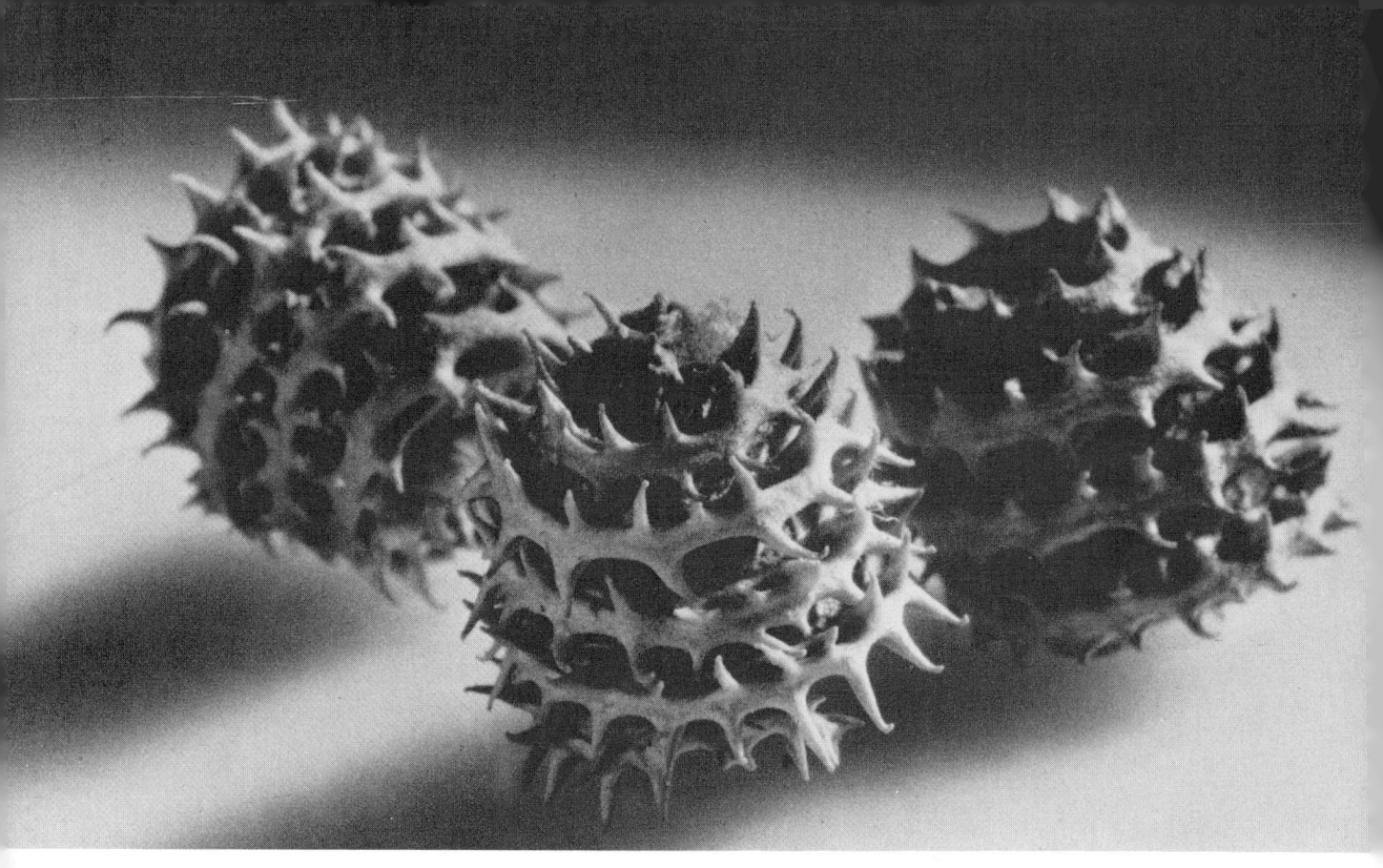

151 · MEDICAGO OBSCURA

The *Medicago* genus, which has about 65 species, is used for forage. Alfalfa, Lucerne and black medick or hop-clover, have very singular fruit. Illustration 151 is the fruit of *Medicago obscura*, found in southern Portugal. The genus belongs to the Leguminosae, the pea family, although there is not much similarity between this fruit and the ordinary pea pod. Here the pod is enclosed in a spiral of four turns, tapering upward and armed with spines, some of which have hooked points. If the base were enlarged, one could see that the spiral begins with a fearsome dragon-like head and a scaly body.

Illustration 153 shows fruit also found in southern Portugal, of another species of the same genus, *Medicago disciformis*. The prickles are longer, but the spirals are otherwise exactly the same. The English species of *Medicago* also have curious fruit as can be seen from the specimen of *M. sativa*, alfalfa, with the whorled, shell-like spirals, shown in Illustration 152.

144

152 · MEDICAGO SATIVA

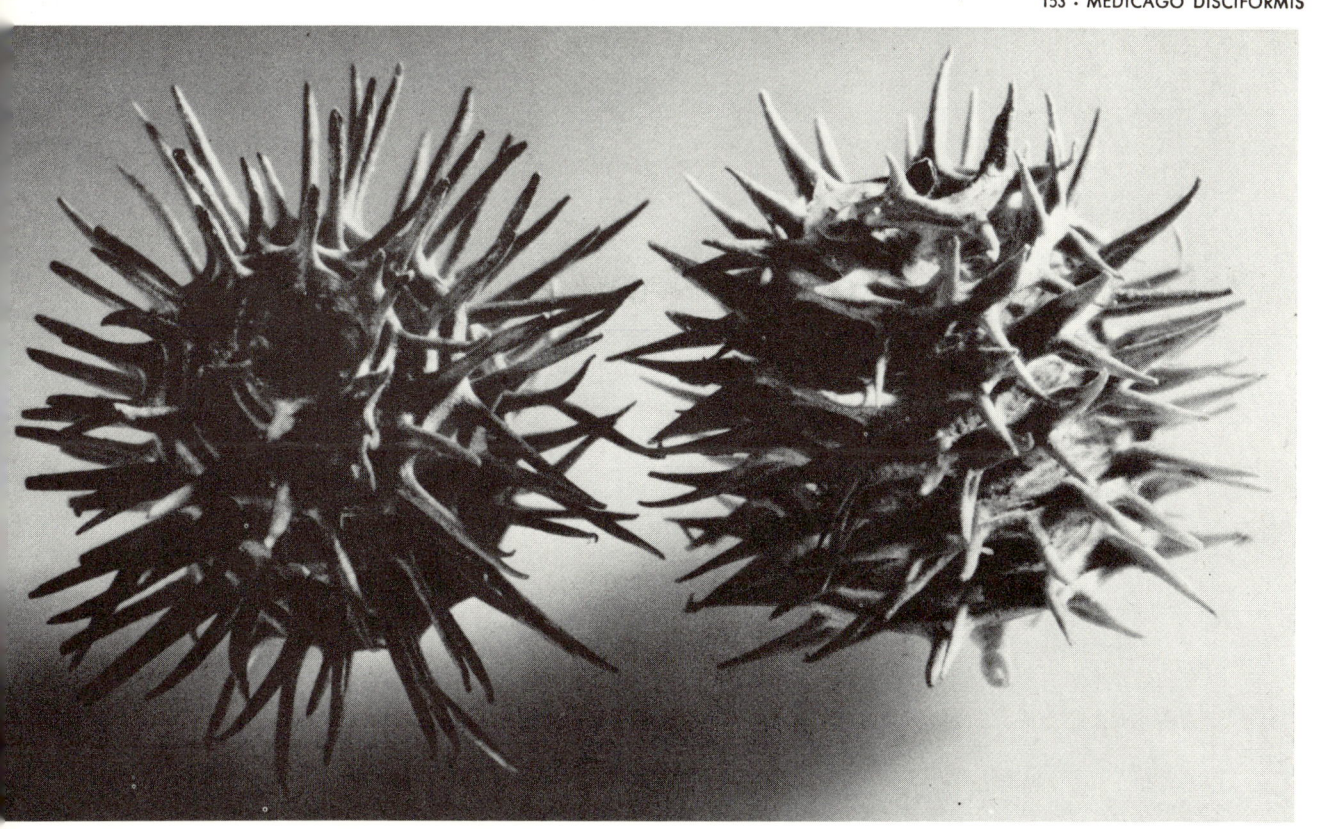

153 · MEDICAGO DISCIFORMIS

145

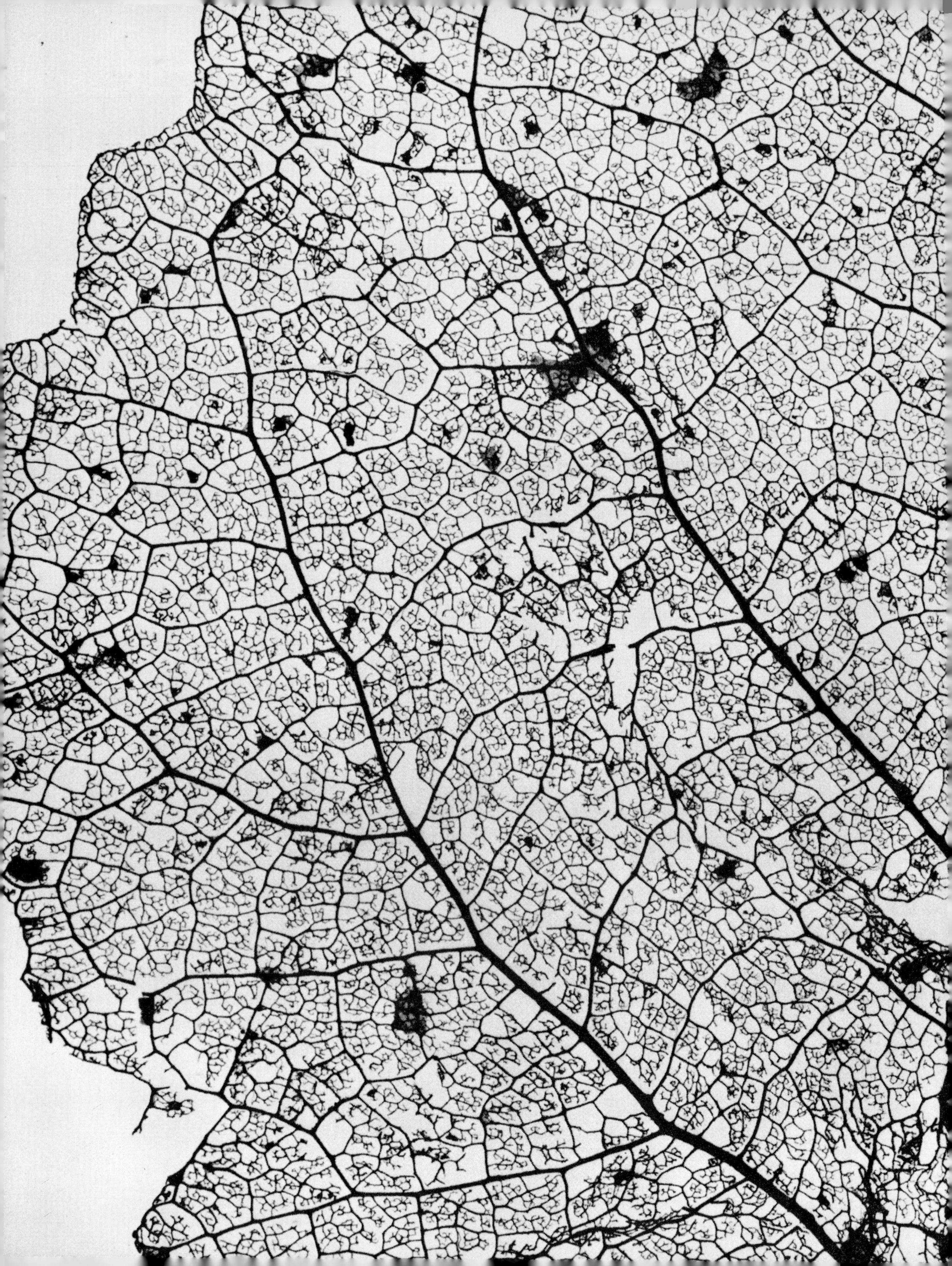

155 · ALCHEMILLA VULGARIS

When leaves fall on dry ground in autumn, they become wrinkled and roll up, but when the trees stand in damp places or have a soft carpet of moss around them, their leaves keep their shape but soon decay. The green color quickly gives way to brown and gray. The nerves of the leaves are most resistant to decay, and even the finest of them survives for quite a long time, while the tissues between them fall away. This leaves a fine network, as seen opposite in the potograph of a skeletal aspen, or poplar leaf.

Most of us admire open leaves, but their development, the various phases between bud and final shape, does not attract us in the same way. The edges of a leaf in bud may be rolled inward or, at times, outward. Some leaves are tightly twisted in spirals, others are folded like closed fans. The leaves of the lady's mantle (*Alchemilla vulgaris*) above show these folds in two stages, the silky, white hairs on the underside can be clearly seen. The shape of the leaves has given the plant its popular name. They are also called dew-cups or my lady's fingerbowl because of the drops of crystal-clear dew that collect in the center of the leaf. According to the botanist, Lindman, these drops are not dew in the ordinary sense of the term, but drops of liquid produced by the plant itself, excreted through the fine pores at the tips of the leaves. The Latin name of the genus refers to the interest shown by alchemists in the drops of dew, as mysterious as the fine drops on the insect-consuming leaves of the round-leaved sun-dew.

154 · POPULUS TREMULA

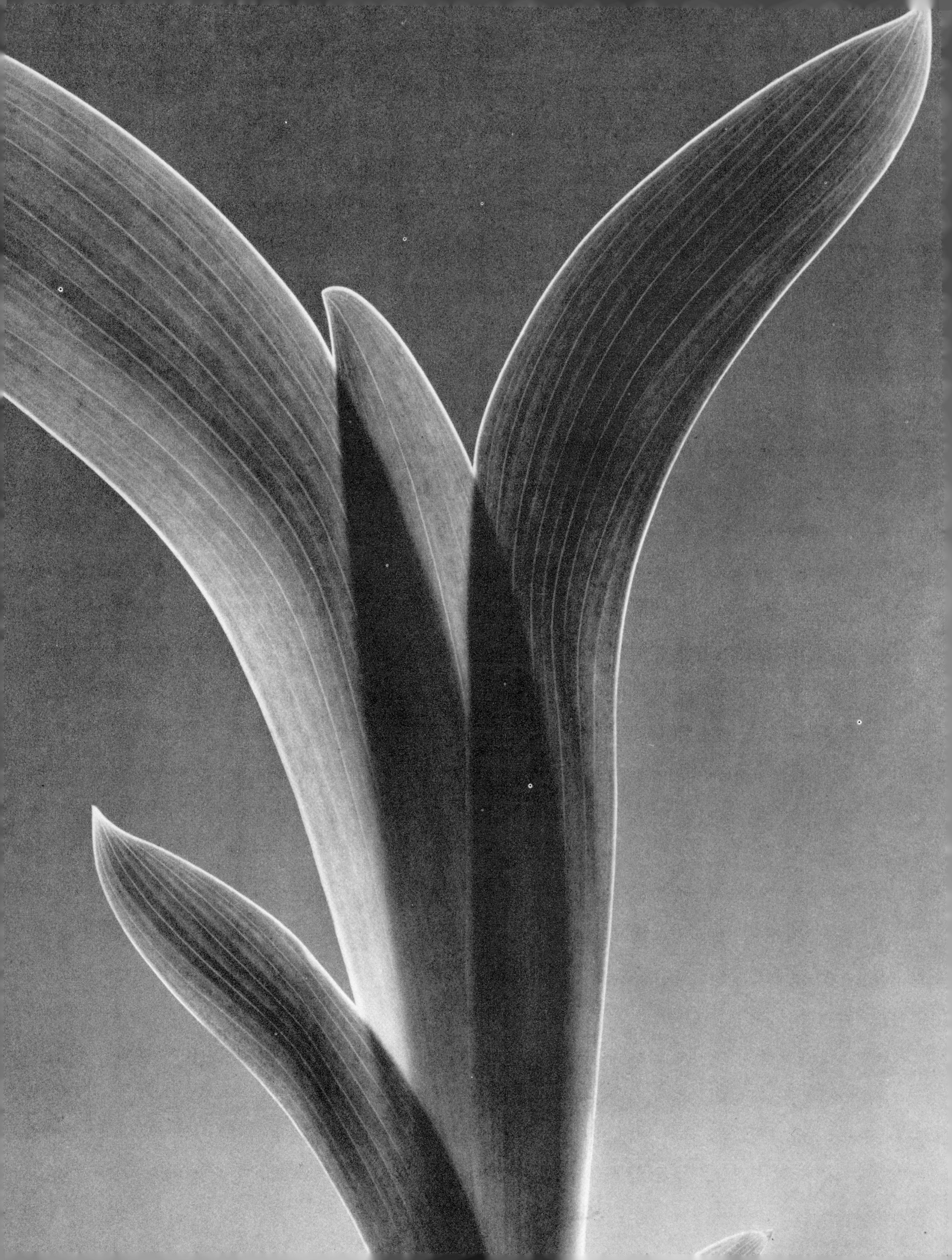

156 - IRIS GERMANICA

*Iris pseudacorus*, the yellow iris, is the only wild species of Iris I have seen. It is called sword-lily because of the long, graceful leaves. A species often cultivated, the *Iris germanica*, has leaves more reminiscent of an oriental scimitar than of a sword. The lines of the leaf nerves and the semi-transparent tissues serve to emphasize its elegant design.

**158 · DROSERA ROTUNDIFOLIA**

Round-leaved sun-dew *(Drosera rotundifolia)* is one of the few carnivorous plants of northern Europe and lives on insects, flourishing in moist areas. The sundew has the same shades of green and red as the white moss in banks of which it grows, together with cranberries, bog myrtle, *Andromeda Polifolia*, and common cottongrass. Insects are caught by the leaves, which are barely one-quarter inch in diameter. The edges and upper sides of the leaves are equipped with a dense border of short red hairs terminating in little knobs. The upper sides of the leaves also have a hairy surface. If a leaf is examined against the light, the border hairs look like a halo with a glistening drop of sticky liquid at the tip of each filament which catches the unsuspecting insect. The contact of an insect on the leaf triggers the mechanism of the trap, the edges of the leaf fold upward and the hairs bend over to hold fast the victim which is finally enveloped and consumed by the plant. When the meal is over, the leaf reopens, the hairs stand upright, and the trap is again ready. If you are interested in following this drama at home, the sundew plant will thrive indoors if it is kept moist and some small insects are collected to feed it.

159

160

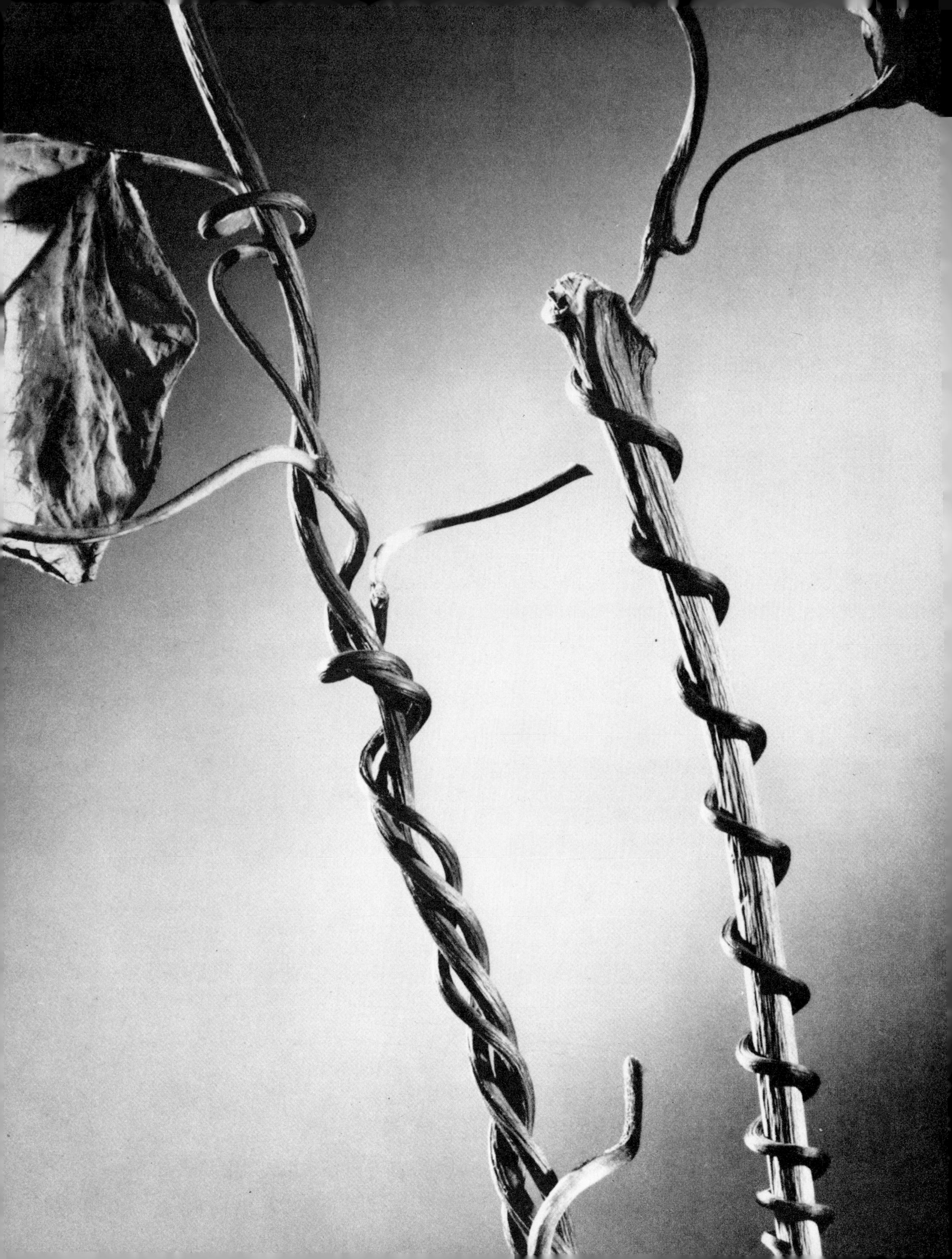

161 · ARISTOLOCHIA DURIOR

There are many different types of movement in plants. The dandelion and windflower close their flowers at sunset, and open them when the light returns. The leaves of the sundew close around the insects that settle on them, and open when they have consumed their victims. Mimosa *(M. pudica)* leaves droop at the slightest touch. Many plants must have some kind of support to reach their full height, other plants require the surface of a wall, cliff, pole or the like. The stem, by winding itself round its support, will climb upward, there are in some plants, special climbing organs which hold the plant upright on a wall, by means of sharp hooks, as the hop, by long, spirally, twisted threads, pumpkins and runnerbeans, for example, or by small, thread-like roots or suckers. Only one climbing plant is presented here, the Dutchman's pipe, *Aristolochia durior (sipho).* In one picture, the thick stem has wound itself around a straight section of another stem, the other shows younger stems twining around each other. The movement is counter-clockwise, as in the runner-bean, for example. Other plants, the honeysuckle and hop, for instance, twine clock-wise. The twining motion is congenital; every plant has its absolutely constant type.

153

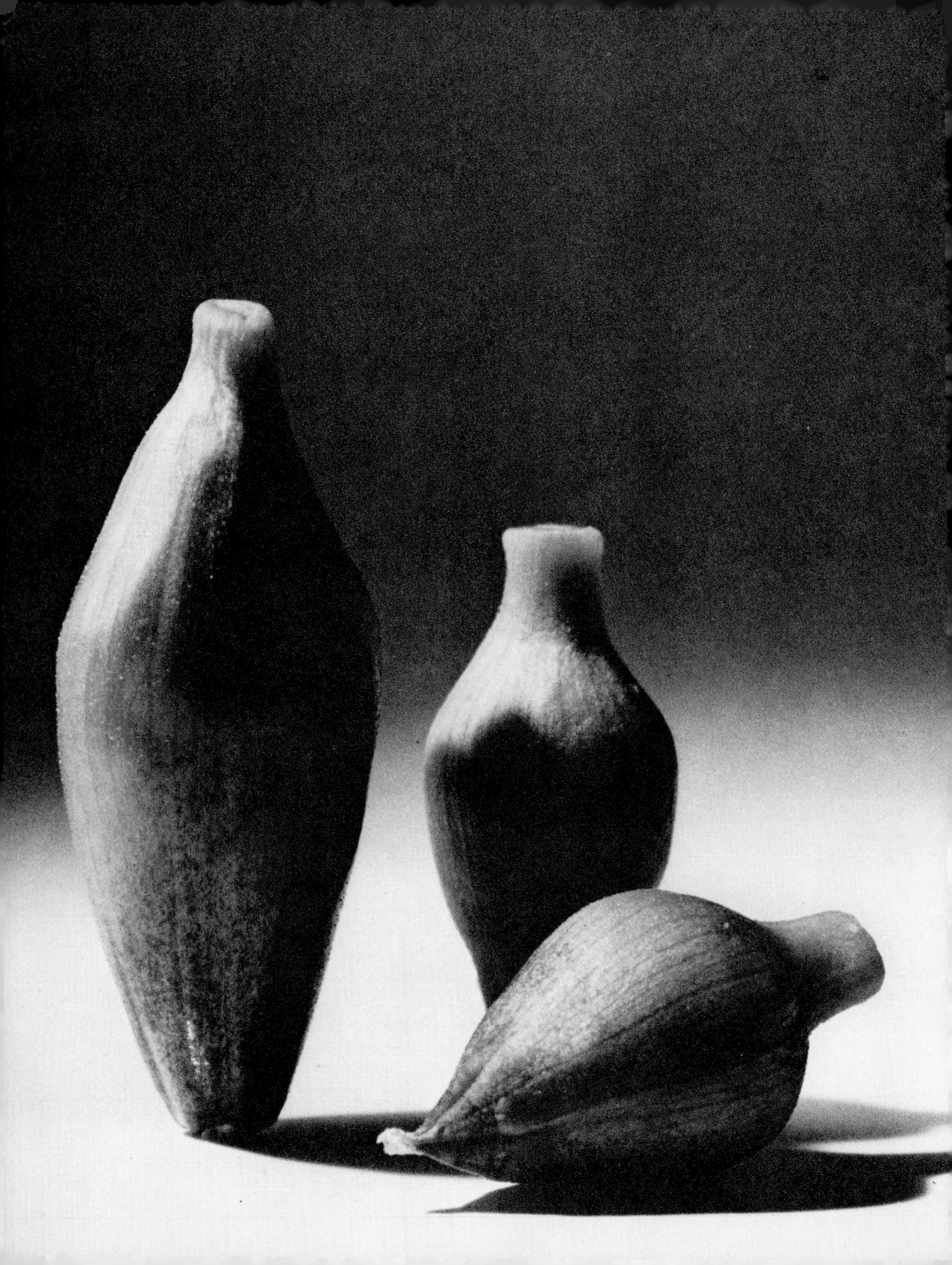

164 · ANEMONE SILVESTRIS

Bouncing Bet of the soapwort *(Saponaria officinalis)* genus, is a cultivated plant, and is found most frequently in old gardens, from whence it often escapes and grows wild in the surrounding areas. The flowers are pale red or white, beautiful at first, but the dense heads soon wither and become untidy and ugly, particularly in wet weather. The buds are the most beautiful part of the plant. Some enlarged buds are shown on the opposite page. They are simple in design, but unusually pure in style, reminiscent of a Grecian amphora. If a few of the withered inflorescences are allowed to remain on the plant to mature, the most beautiful tiny seedcases will be found, well hidden under the remnants of the wrinkled sepals. But the stem should not be overlooked. It gets harder and harder as time goes by, until it is wooden, with distinct joints, astonishingly like bamboo.

*Anemone silvestris*, the snowdrop anemone, is grown for its large white flowers, but it must be kept under control, for it easily spreads in all directions with the vitality of a weed. The buds hang down like heavy drops, but turn up as they open. This movement of the flower stalk is a very elegant one, and it is interesting to see how the individual flowers vary the theme.

**165 · ALLIUM SCHOENOPRASUM**

The culinary features of chives (*Allium schoenoprasum*) are widely appreciated. Only the tubular leaves are used in the kitchen, thus there are very few housewives who have ever seen chive blossoms although the flowers are well worth seeing. They form a crowded group at the top of a naked stem; as buds they are enveloped in a thin membrane which keeps the florets together in a bulb-shape cupola. When the membrane bursts, the buds appear to be a collection of elegant vases or flasks in pale green and violet, patterned with darker stripes. When full-blown, the six-lobed chalices are, as is usual in the lily family to which this plant belongs, perfect in form.

Illustration 166 shows a closed pod and one which has just opened. They are the pods of *Pithecolobium dulce*, from Santos, Brazil, but may be found in other tropical areas of South America. The white seeds stand out against the brownish-red surface when the pod opens in a handsome whorl. It is popular for the taste of the pulpy tissue surrounding the unripe seeds. The word *dulce*, really implies sweet in the sense of savoury, but here it may be said that the fruit is not only pleasing to the palate, but also to the eye. The fruits are probably eaten by apes as indicated by the name, *Pithecolobium*, which means ape-pods.

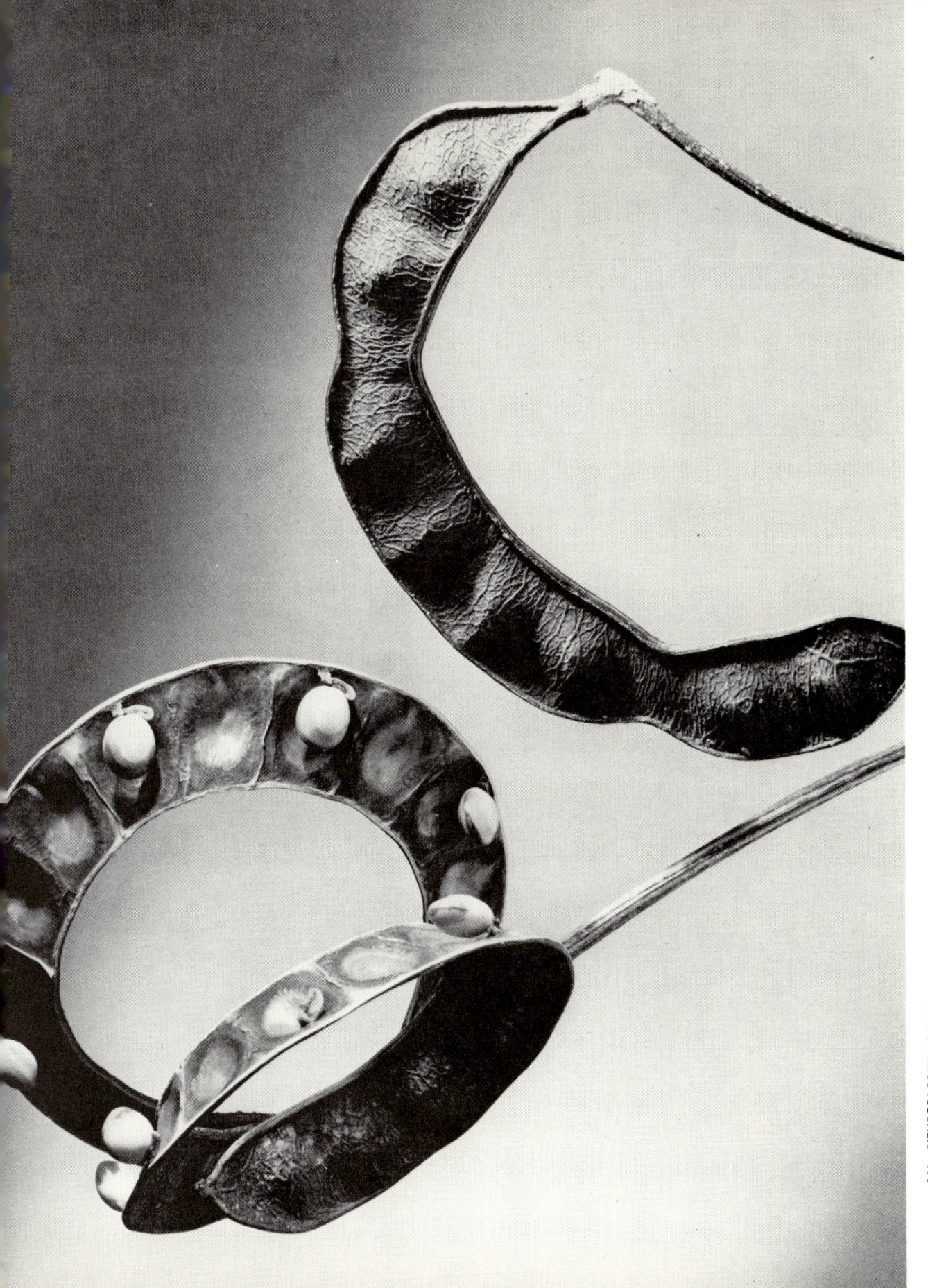

166 · PITHECOLOBIUM DULCE

167 · CYTISUS LABURNUM

Laburnum (*Cytisus laburnum*) needs no introduction; it is well known and loved by all who are interested in plants. Most of the laburnum trees and bushes in Sweden were killed during the severe winters of the early 1940's, but they are now reappearing. Even after the glorious yellow flowers have withered, the laburnum is beautiful, with its pale green foliage, its green branches and twigs, and particularly the young shoots which are covered with silvery hair.

The gray-brown clusters of pods usually remain throughout the winter, and since the laburnum fruits very profusely, the bushes look quite untidy at that time. But when the pods begin to open, the laburnum gives us a new experience of beauty, if we take the trouble to look closely. As is usual with plants belonging to the pea family, the pods split into two halves and twist apart in spirals. The coal-black seeds contrast beautifully with the pale inside of the pods.

168 · LYCOPODIUM CLAVATUM

Of the nearly 200 species of *Lycopodium* known to exist, there are only about a half-dozen in England. The one shown here is *Lycopodium clavatum*, common clubmoss, or stag's-horn moss. The pale green cones rise in small colonies from the recumbent trailers. If the ripe clubs are struck, a cloud of spores rises from them. Formerly, these spores were collected and sold to chemists, who used them to prevent pills from adhering to each other. The Greek word *Lycopodium* means wolf-paw, and in an old book dating from the late seventeenth century, the plant is called wolf's-foot. In Germany, the name Bärlapp (bear's paw) occurs, and such names as Devil's foot have also been reported. These names probably arose from an imagined likeness between the rough trailers or the cones, and the paws of various wild creatures. Mats made from the trailers, are responsible for the common name of the species, meaning mat-moss or mat-grass.

169 · EQUISETUM ARVENSE

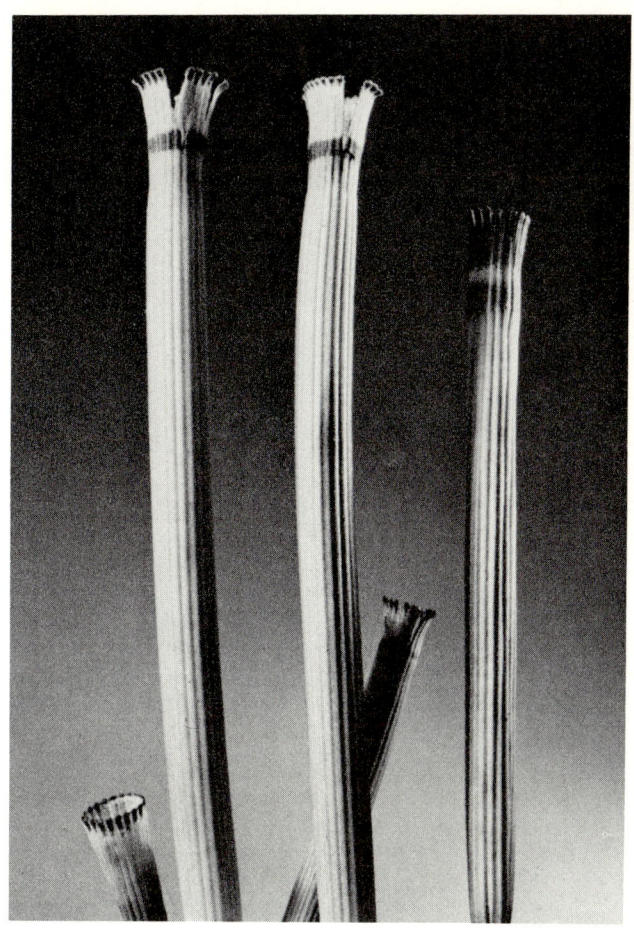

170 · EQUISETUM LIMOSUM

171 · EQUISETUM ARVENSE

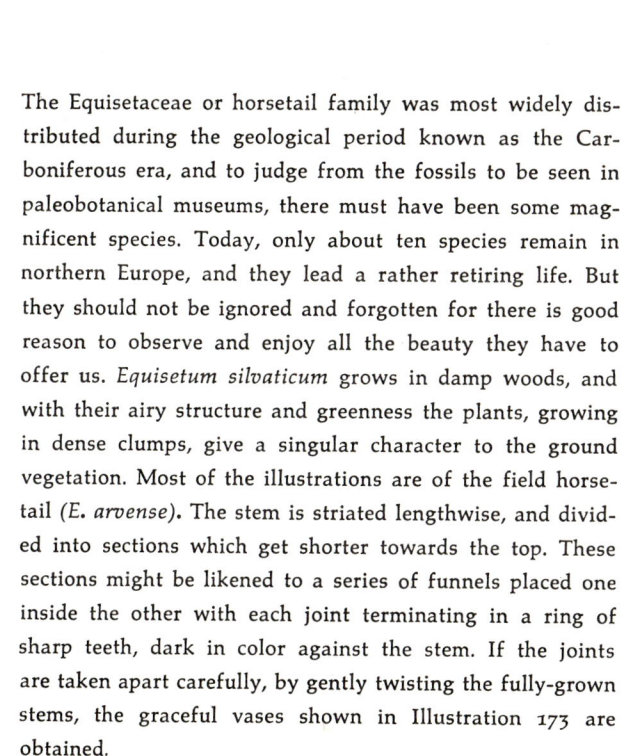

The Equisetaceae or horsetail family was most widely distributed during the geological period known as the Carboniferous era, and to judge from the fossils to be seen in paleobotanical museums, there must have been some magnificent species. Today, only about ten species remain in northern Europe, and they lead a rather retiring life. But they should not be ignored and forgotten for there is good reason to observe and enjoy all the beauty they have to offer us. *Equisetum silvaticum* grows in damp woods, and with their airy structure and greenness the plants, growing in dense clumps, give a singular character to the ground vegetation. Most of the illustrations are of the field horsetail *(E. arvense)*. The stem is striated lengthwise, and divided into sections which get shorter towards the top. These sections might be likened to a series of funnels placed one inside the other with each joint terminating in a ring of sharp teeth, dark in color against the stem. If the joints are taken apart carefully, by gently twisting the fully-grown stems, the graceful vases shown in Illustration 173 are obtained.

The inflorescence (cone with spore-sacs), which appears early in spring before the leaves have opened, is of a curious construction in the bud stage, having a mosaic-like structure. When the pieces of mosaic separate from each other, the bluish-green pollen pours out in clouds. If you have a magnifying glass handy, use it. The pollen is a fantastic sight; the entire field is in an uproar, a mass of tiny balls, milling round and round like a shoal of spawning fish. Each tiny ball is a spore, and the motion begins when the spore membrane bursts into four narrow strips forming "elators," which scatter the spores by their hygroscopic movement, a result of their rapid drying.

Another species is also shown, *E. limosum*, the water horsetail (Illustrations 170 and 172). It is more delicate in structure, with long striated stems. The sheath terminating each section of the stem is white, with a black ring at the base. The sharp teeth on the sheath soon fall away, leaving the square-cut edge, patterned in black.

172 · EQUISETUM LIMOSUM

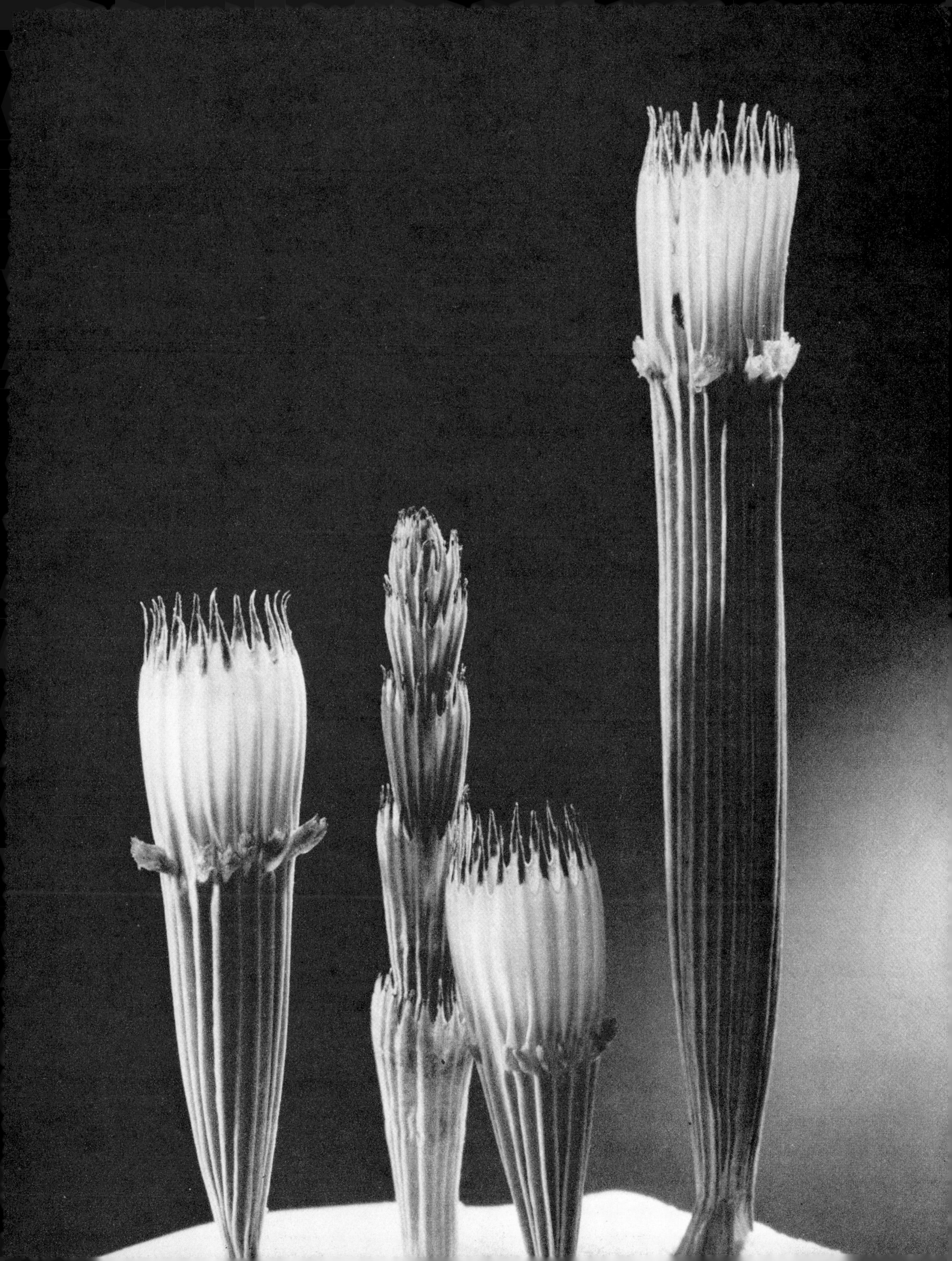

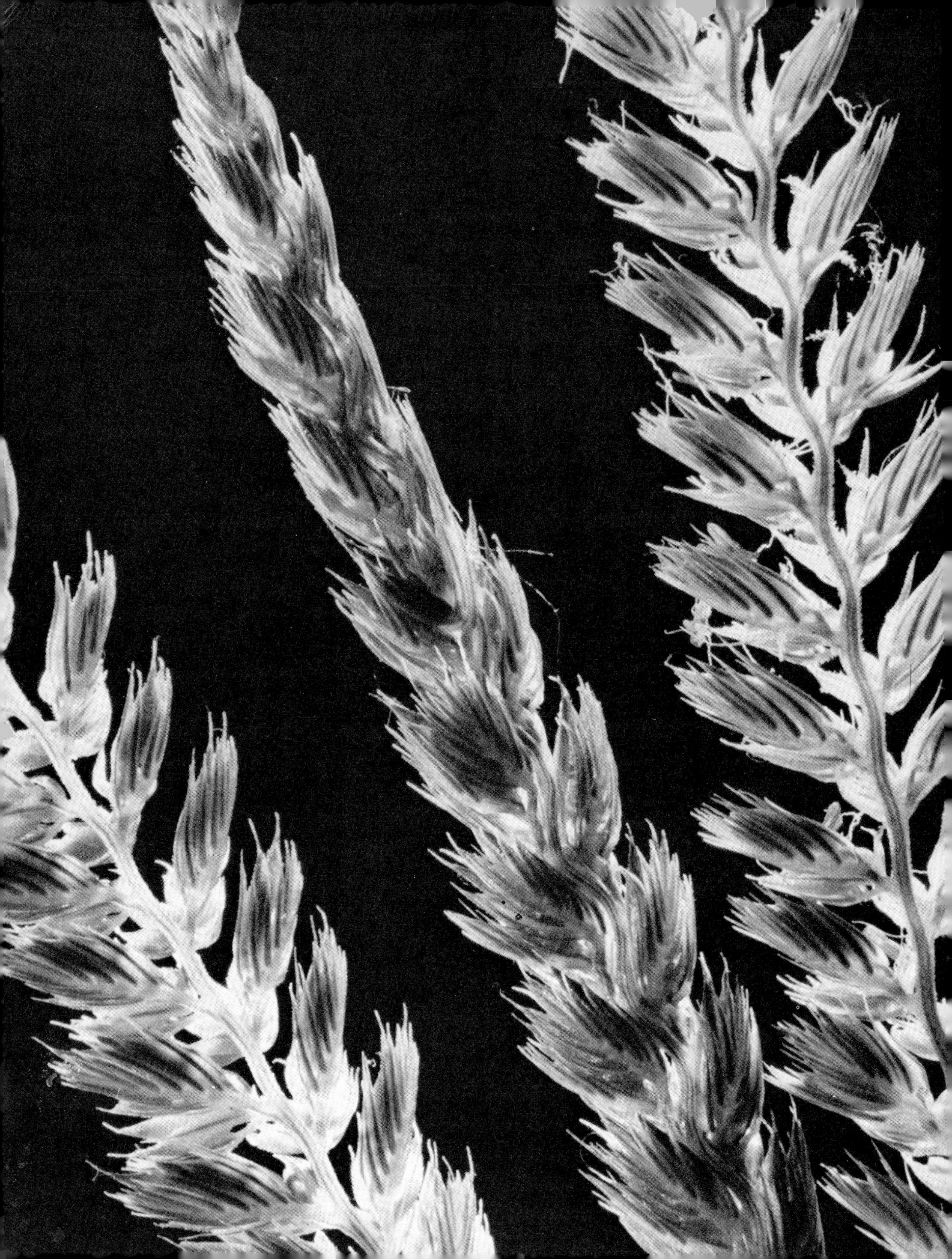

174 · PHLEUM PRATENSE AND CYNOSURUS CRISTATUS

When one considers how large the family of grasses is, with thousands of species in the world—even a short flora usually lists at least a hundred—it is surprising how few of them most people can identify. For many, grass is a term which reminds them of lawns or pastures, or just a tall stem with a tuft at the top. The most common grasses—wheat, rye, barley, oats and the reeds around lakes—are hardly counted as grass; nor is the timothy in fields. Apart from these, genuine grasses have very few popular names, and similar species often go under the same name. Linnaeus tried to assign names to specific species, but even in standard floras the nomenclature used in the description of grasses is extremely confusing. But it is not necessary to know either Latin or popular names to enjoy the grass family.

It has so much beauty to offer us—the greenness, the waving stems, the airy plumes, the curiously constructed flowers with their ingenious details, particularly the stamens, with their anthers swaying in the gentlest breeze. Calder, the American artist, and his followers are novices compared with Nature in "Mobile art." The illustration opposite shows spikes of crested dog's tail *(Cynosurus cristatus)*. The wavy line of the stem and the branched leaflets, which support and protect the stamens and pistils can be seen clearly. In the photograph above these leaflets are isolated. The characteristic leaflets of timothy *(Phleum pratense)*, have a tiny lyre at the top, which gives the spike its familiar roughness.

165

175 · CYNOSURUS CRISTATUS

The sedges, called *Carex* in Latin, are not among the plants whose flowers attract attention. They have no bright colors, and the general appearance of the plant is most insignificant and humble, but on closer examination, as is usually the case when one studies anything in nature carefully, novel and interesting designs can be found. Here is an example of the bladder-sedge *(Carex vesicaria)*, one of the most common species, often found in company with marsh cinquefoil and *Lysimachia thrysiflora.* In early summer the spikes stand upright, but as they ripen they bend in graceful arches under the weight of the seeds. The seedcases are close together in rows along the stem of the spike, pale green in color, inflated, and look like beautifully designed bottles decorated with faint, longitudinal lines.

The inflated seedcase is of prime importance in the dispersion of the seeds, for the empty space serves as a buoy which can keep the seed afloat as it is carried away from the parent plant on the water.

176 · CAREX VESICARIA

178 · CLADONIA GRACILIS VAR. CHORDALIS

When one's eyes have been opened to the fascinating shapes of goblet lichen, drawn by the tempting variety, it is easy to get lost in the woods. In a rocky country region with occasional pine and spruce, dense carpets of these lichens may be found, and patch after patch of different kinds of lichen *(Cladonia)* are discovered. One never tires of studying the infinite variety of shapes nature has created, and from intently peering downward while walking, one suddenly may find oneself lost. When modern sculptors display their work in art galleries, we are given the opportunity of seeing all kinds of new creations, distorted representations of the human body; thread-like figures; shapeless and gigantic limbs; laboured motion; abstract compositions and mobiles. Art critics often stand confounded when confronted with these works, and their indecision and confusion can be sensed in their reviews. As a layman in this field, I can only admit that I occasionally feel as if I understood the intention of this art and I enjoy the contrast of line and surface although I cannot recognize any similarity or relation to pre-formed creations, therefore these sculptures are more artificial than art to me. It is quite different with the discoveries made in nature. Everything is accepted as purposeful, and nothing is ugly.

*C. pyxidata*, pale goblet lichen, has a pronounced horn shape. It frequently grows in colonies of about a dozen plants, and, in spite of its small size, it gives a monumental impression, due to its austere and simple design. At times the bent horns look like tiny trombones in an orchestra.

179 · CLADONIA GRACILIS

180 · CLADONIA PYXIDATA

The stems of the goblet lichens are called podetia by botanists, and their spore-producing organs, apothecia. In the *Cladonia* species, the apothecia are irregular club-shape swellings at the top of the podetia or their branches. Some species have deep red fruit organs, and it is a splendid sight that confronts us when we find colonies of these lichens growing among miniature forests of grayish-white reindeer-lichen on a rock or a rotting tree stump. One of the illustrations shows the species, *Cladonia gracilis*, with brown wart-like fruit organs on most of the stems.

Moss, in everyday speech, is a collective term; one thinks of the carpets and banks of moss in bogs, or the soft blankets spreading over stones and tree stumps, which give woods an atmosphere of primeval wilderness. Mossy is a descriptive adjective. With mosses, as with so much in the world of plants, much beauty is overlooked unless details are examined. The branched stems with their fine leaves are most beautiful in spring and early summer, when the young buds stand out pale green against the older, darker growth. Illustration 181 is the house-moss (*Hylocomium splendens*) with slender bud-shoots, which grow upward with arched tops, exactly like a cobra raising its head to strike. The Latin name indicates that it is a luminous moss which prefers woods. The second photograph (Illustration 182), shows the stems and leaves of one of our most beautiful mosses (*Ptilium crista-castrensis*), feather-like in structure, and called comb-moss. It was gathered in spring, from a damp, wooded slope growing in large carpets together with scattered colonies of windflowers (*Anemone nemorosa*).

181 · HYLOCOMIUM SPLENDENS

182 · PTILIUM CRISTA-CASTRENSIS

The seedcases of mosses are more interesting than stems and branches when studied in detail. The sporogonia or spore-sacs, rise on slender stems above the foliage. Nature experiments with form in many ways. The plants may grow to a height of a foot, and the sporogonia, which often occur in large numbers, rise high above the green tussocks on strong, slender, elegant stems. At first the sporogonium is covered with a pale, pointed cap which loosens easily when pulled by the tip. When the seeds ripen, the lid of this tiny goblet can be lifted by its knob, and using a magnifying glass to continue your examination, you will find that the upper, square-cut surface of the sporogonium is like a fine sieve, fastened around a narrow fillet. When the spores are ripe, they sift through the tiny holes if the sporogonium is turned upside-down. They would make excellent Lilliputian powder-boxes.

Illustration 183 shows *Polytrichum formosum* with the caps still on the sporogonia. The generic name means very hairy, for which the ragged caps are obviously responsible, and cause the sporogonium to look like birds on a pole.

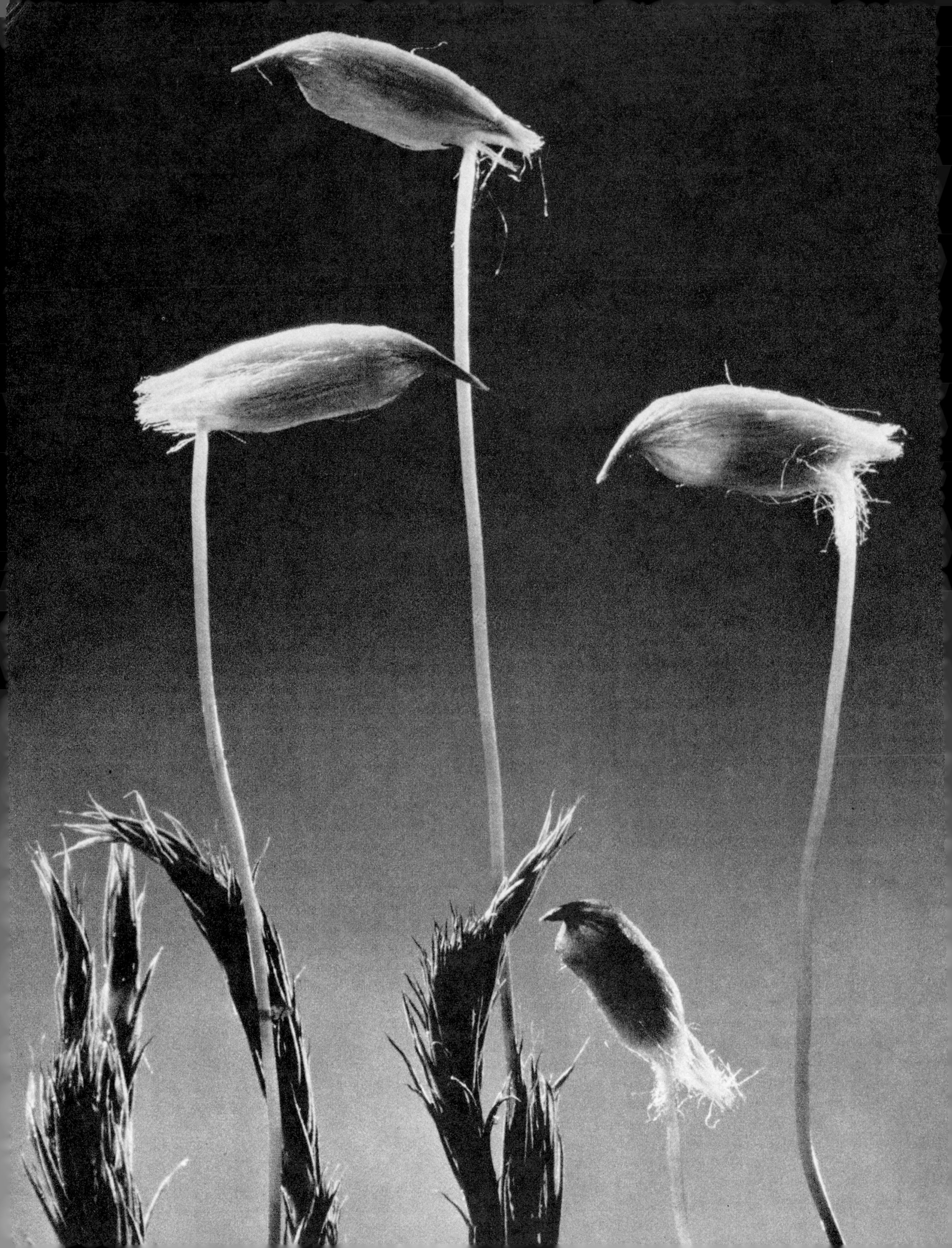

184 · SOLANUM TUBEROSUM

Towards the end of the winter, when picking over potatoes in the storage bin, many of them are found to have become soft and wrinkled, with shoots growing from the eyes. At first these shoots are green, but become paler as they grow and they may reach a yard in length. In a potato cellar that has been neglected for some time, a forest of shoots, long and pale, slender and weak, may be found striving longingly upward from their dark dungeon, instinctively seeking the light. But the small shoots, an inch long are the most interesting as they vary infinitely and attain the most imaginative shapes. When I was a child I liked to lie on my back in summer to watch the drifting clouds form giants, elephants, tigers, all kinds of wonderful things in the sky—indeed, I still enjoy this whimsy and it is almost the same with new potato shoots. The illustration above suggests some prehistoric lizards gazing out over a deserted landscape.

171

183 · POLYTRICHUM FORMOSUM

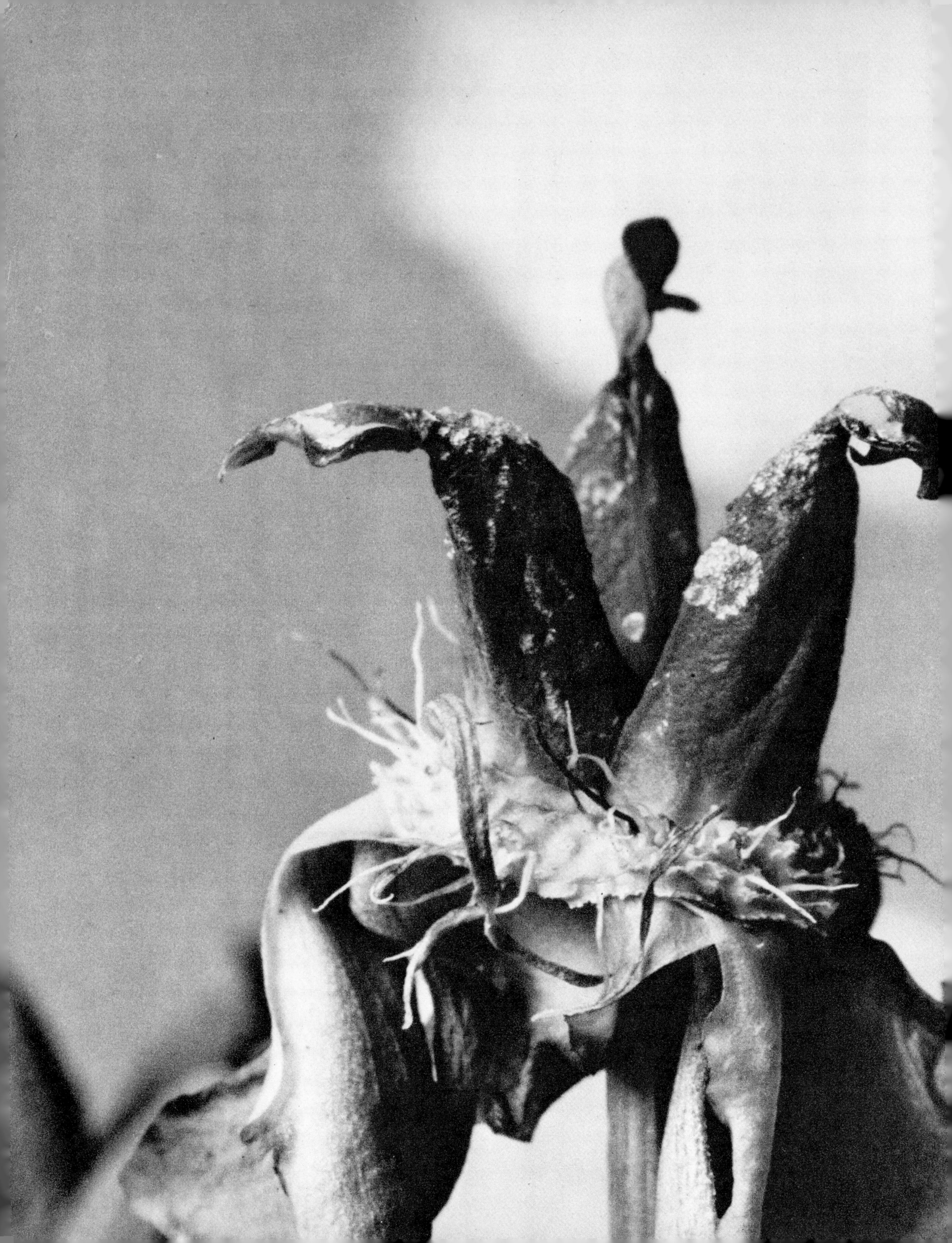

186

187 · PAEONIA SUFFRUTICOSA

There cannot be many people who have never seen a peony (*Paeonia officinalis*), or some variety of it in someone's garden. In beauty, the peony's enormous red, white, or pink flowers can challenge any rose. They fade in beauty as the petals wither looking healthy and fresh, but at that stage it is time to clean up in the garden; the petals must be swept up and removed and the stalks cut down. But the gardener interested in the plant world should let some peonies set fruit and follow the development, week after week. After the petals fall, the dense cluster of stamens withers, and then there remains only the receptacle, a pale red disc from which the rudiments of the fruit rise. These vary in number from two to four or more, and look like a small family of birds in a nest. The stigmas form the birds' heads. They remind one of penguins, sometimes strongly of vultures. The appearance and attitude of the stigmas vary greatly, giving life and movement to the birds. They may stretch their heads upward on long necks, or now turn them aside, or preen their feathers with their bills. As the days pass by, the fruit grows, and the birds get fatter and more dignified; their bodies often are striped lengthwise, unless they are covered with down, and their red color turns green and later, brown. One cannot help stopping to look at these curious "birds."

Eventually, the birds become shapelessly fat, lose their elegant figures and their heads. The fruits continue to ripen, bend away from each other, until they lie outstretched, forming three- or four-pointed stars. The walls of the fruit harden and burst open on the upper side, revealing the large seeds. The fissure widens, the seeds fall out, and there remains a star of empty bowls, striated inside.

185 · PAEONIA OFFICINALIS

The most descriptive English name of *Antirrhinum* is snapdragon, and the present fashion of calling the snapdragon Antirrhinum is to be deplored. It is well named, for the flower really looks like an animal's wide jaws, gaping hungrily when the sides are pressed together. Other old names are calves-snout and lion's snap.

As the inflorescence grows and one flower after another bursts into bloom, the lower flowers wither and fruit-setting begins a strange little zoo of dragons, chicks and monsters. God probably smiled with pleasure while designing this plant. First the curious flower, then the half-ripe fruit strikingly like a row of baby chicks sitting along a perch, and finally the ripe fruit frightening to behold, like veritable monsters with wide-open mouths and dark eyesockets from which the tiny seeds pour.

Edward Lear, the English humorist who popularized the limerick in England with his *Book of Nonsense*, had, among his amusing pictures, a series called "Nonsense Botany." The snapdragon with its flowers and fruit would not have been out of place there. It is very similar to Lear's imaginary plants and he probably would have called it *Leopulla diabola*.

188 · ANTIRRHINUM MAJUS

189

190 · SCORPIURUS SULCATUS

I found the pale specimens of the fruit shown above, *Scorpiurus sulcatus*, in the province of Algarve, in southern Portugal. There were no flowers to be seen, as they had withered before I arrived at the beginning of June, but I know they are something like our bird's-foot trefoil. *Scorpiurus* means scorpion's tail. I don't think this an appropriate name; for with fruit like this the plant should be renamed *Serpens spinosa*, the prickly snake, and I hope that at their next nomenclature congress botanists will change the name. There are good reasons for the species name, *sulcatus*, for the belly of the snake is longitudinally striated, but the spines on the ridges between the furrows on the back dominate the picture. The head is thrust forward with its pointed snout, and there is often a crest at the back of the head, the remains of the calyx. We often say, as alike as two peas, but this does not hold true of *Scorpiurus* fruit. They vary infinitely; sometimes the snake is curled up, with its head resting as if asleep, at others the head is lifted as if ready to strike, and often it appears as if it were attentively listening to a snake-charmer's flute.

191 · CITRUS AURANTIUM

I seem to remember that in my youth, oranges had many more pips than they now have and they were not as flavorful as they are today. But we must not be annoyed with the pips we do find for on close examination they can be quite amusing. They vary greatly in shape, and are very much like birds, with a distinct head, spindle-shape body and a short tail. Impaled, as in this whimsical illustration, on the sharp spines of a blackthorn bush, for example, the similarity to birds perched on a branch is striking.